The Cambridge Diet
A Manual for Practitioners

The Cambridge Diet

A Manual for Practitioners

by

John Marks MA, MD, FRCP
Fellow, Tutor and Director of
Medical Studies, Girton College,
Cambridge, UK

and

Alan Howard MA, PhD, FRIC
Lecturer in Nutritional Research,
Department of Medicine, University of Cambridge,
Cambridge, UK

MTP PRESS LIMITED
a member of the KLUWER ACADEMIC PUBLISHERS GROUP
LANCASTER / BOSTON / THE HAGUE / DORDRECHT

Published in the UK and Europe by
MTP Press Limited
Falcon House
Lancaster, England

British Library Cataloguing in Publication Data

Marks, John, *1924–*
    The Cambridge diet: a manual for
    practitioners.
    1. Reducing diets   2. Physical fitness
    I. Title   II. Howard, Alan N.
    613.2'5   RM222.2
ISBN 978-94-011-8013-9   ISBN 978-94-011-8011-5 (eBook)
DOI 10.1007/978-94-011-8011-5

Published in the USA by
MTP Press
A division of Kluwer Academic Publishers
101 Philip Drive
Norwell, MA 02061, USA

Frome and London

Contents

Foreword

Obesity in affluent countries continues to be a serious problem. When one runs an Obesity Clinic there appears to be an unending series of patients who need help.

Our understanding of weight control has been advanced not only by the studies of obese patients but also by our investigation of the problems of weight gain in young women with anorexia nervosa. Just as among obese patients there is the group of 'efficient metabolizers' who can maintain their excessive weight with a calorie intake occasionally as low as 600 kcal per day, so we have demonstrated that among anorexic women there are a few who may fail to gain normal weight with 3500 kcal per day. Some of the latter group may in fact develop $\bar{T}_3$ (tri-iodothyronine) thyrotoxicosis.

One of the major unknown factors is still what determines when patients may be at these extremes. If we knew how to control these extremes we would like, for a time, to switch each to the opposite end of the spectrum. One factor which is becoming apparent from studying people in the machine which continually plots metabolic rate is that the metabolic response to food is a major factor in determining a person's weight. In general terms there is a tendency for those who are below their ideal weight to have an increased metabolic response to food and those who are above ideal weight to have a reduced metabolic response to food.

The normal individual is able to vary his food intake over a considerable range and yet his/her weight stays surprisingly constant. The anorexic young woman has a very poor weight response to increased food intake, whereas most overweight people have a disproportionate weight gain with relatively modest increases in food.

Having said that, it is nevertheless true that a considerable number of obese patients have achieved that state by excessive ingestion of calories. Many of them can in fact achieve normal weight by consistent adherence to a food intake around 1000 kcal. Only those who drop their metabolism too much during dieting have to use more severe dieting methods. Our extensive experience of the diet based on milk

protein, lactose, extra vitamins and minerals has shown that this is a safe and effective dieting procedure. The preparation of the Cambridge Diet has made it much easier for people to use this type of diet.

There is, however, another group in whom the liquid only diet is essential. These are the people who have some degree of compulsive eating. This varies from those who have phases, of various lengths, when they cannot stop nibbling when under pressure, to those who have the extreme degree of bingeing seen in the young people with bulimia. For many of these the liquid only diet may be the only effective way to control their bingeing, though getting through the first week is always extremely difficult and sometimes impossible. For them, the return to solid food may again face them with difficult problems.

For those who have lost their obesity by dieting, the major problem is always subsequently restricting their calorie intake to keep it in balance with their metabolic use of energy. All too often people think they can return to their previous eating habits and inevitably gain excessive weight. Education in sensible eating to maintain normal weight is extremely important. Since many people diet without consulting their doctors, it might be that the Counsellors trained by Cambridge Nutrition might add to the body of informed opinion to help in the education of the public in eating for more healthy living.

Ivor H. Mills, PhD, MD, FRCP, Hon FACP
Professor of Medicine
University of Cambridge
September 1986

Part I
OBESITY

1

Obesity and its problems

'More are slain by suppers than the sword' – Old proverb[1]

When a practitioner is faced with a patient who is grossly obese there is no difficulty in defining and recognizing the condition. Moreover it is rare for the patient to deny that a problem exists, to deny its extent or to agree that treatment is required – though whether successful treatment can be achieved is another matter.

However, for lesser degrees of obesity medical diagnosis and patient acceptance may be more difficult.

Excess weight can be viewed from two different points of view – the objective measurement of weight and the subjective measurement of body image. When a difference exists between the two the practitioner may have difficulty convincing the patient about the need for corrective action.

The main objective determination of the normality of the weight is based on actuarial considerations. That is to say, above (or indeed below) a given range of weights morbidity and mortality increases. One method of expressing this ideal weight range is based upon a proportion of weight for height. The most commonly used formula for this is the body mass or Quetelet index[2]. This is derived by dividing the weight (in kilograms) by the height squared (in metres). Based on this formula the acceptable weight range is between 20 and 25; obesity is taken to start at either 27 or 30 by various authorities; gross obesity at an index of 40. Although the Quetelet index is probably the best of the height–weight indices[3] there are doubts about its value when assessing obesity or fat content in individuals[4].

Another objective method is the table of weight for height. A widely accepted version is that published in 1959 by the Metropolitan Life Insurance Company of New York. A typical table of this type is given in Table 1. Obesity is normally taken as starting at 20% above the upper end of the acceptable range of weight for height. Some workers in the field have also defined an arbitrary level of 60% above the acceptable range as being grossly obese. At his 60% overweight level

3

Table 1 Height and weight tables (desirable weights (in indoor clothing))

Imperial measurements

	Men of ages 25 and over				Women of ages 25 and over		
		Weight (lb)				Weight (lb)	
Height ft in	small frame	medium frame	large frame	Height ft in	small frame	medium frame	large frame
5 1	112–120	118–129	126–141	4 8	92– 98	96–107	104–119
5 2	115–123	121–133	129–144	4 9	94–101	98–110	106–122
5 3	118–126	124–136	132–148	4 10	96–104	101–113	109–125
5 4	121–129	127–139	135–152	4 11	99–107	104–116	112–128
5 5	124–133	130–143	138–156	5 0	102–110	107–119	115–131
5 6	128–137	134–147	142–161	5 1	105–113	110–122	118–134
5 7	132–141	138–152	147–166	5 2	108–116	113–126	121–138
5 8	136–145	142–156	151–170	5 3	111–119	116–130	125–142
5 9	140–150	146–160	155–174	5 4	114–123	120–135	129–146
5 10	144–154	150–165	159–179	5 5	118–127	124–139	133–150
5 11	148–158	154–170	164–184	5 6	122–131	128–143	137–154
6 0	152–162	158–175	168–189	5 7	126–135	132–147	141–158
6 1	156–167	162–180	173–194	5 8	130–140	136–151	145–163
6 2	160–171	167–185	178–199	5 9	134–144	140–155	149–168
6 3	164–175	172–190	182–204	5 10	139–148	144–159	153–173

Metric measurements

	Men of ages 25 and over				Women of ages 25 and over		
		Weight (kg)				Weight (kg)	
Height (cm)	small frame	medium frame	large frame	Height (cm)	small frame	medium frame	large frame
155	50.5–54.5	53.5–58.5	57.0–64.0	142.5	42.0–44.5	43.5–48.5	47.0–54.0
157.5	52.0–56.0	55.0–61.0	58.5–65.0	145	42.5–46.0	44.5–50.0	48.0–55.0
160	53.5–57.0	56.0–62.0	60.0–67.0	147.5	43.5–47.0	46.0–51.0	49.5–56.5
162.5	55.0–58.5	57.5–63.0	61.0–69.0	150	45.0–48.5	47.0–52.5	51.0–58.0
165	56.0–60.0	58.5–65.0	62.5–71.0	152.5	46.5–50.0	48.5–54.0	52.0–59.5
167.5	58.0–62.0	56.0–66.5	64.5–73.0	155	47.5–51.0	50.0–55.0	53.5–61.0
170	60.0–64.0	62.5–69.0	66.5–75.0	157.5	49.0–52.5	51.0–57.0	55.0–62.0
172.5	62.0–66.0	64.5–71.0	68.5–77.0	160	50.5–54.0	52.5–58.5	57.0–64.5
175	63.5–68.0	66.0–72.5	70.0–79.0	162.5	52.0–56.0	54.5–61.0	58.5–66.0
177.5	65.0–70.0	68.0–75.0	72.0–81.0	165	53.5–57.5	56.0–63.0	60.0–68.0
180.5	67.0–71.5	70.0–77.0	74.0–83.5	167.5	55.0–59.5	58.0–65.0	62.0–70.0
183	69.0–73.5	71.5–79.0	76.0–85.5	170	57.0–61.0	60.0–66.5	64.0–71.5
185.5	71.0–76.0	73.5–81.5	78.5–88.0	172.5	58.5–63.5	62.0–68.5	66.0–74.0
188	72.5–77.5	76.0–79.0	80.5–90.5	175	61.0–65.0	63.5–70.0	67.5–76.0
190.5	74.0–79.0	78.0–86.0	82.5–92.5	178	63.0–67.0	65.0–72.0	69.5–78.5

For girls between 18 and 25, subtract 1 lb (454 g) for each year under 25. (Courtesy of Metropolitan Life Insurance Co.)

mortality is at least double that of those with a normal weight. However, it is probably better to use the terms 20%, 30% etc., above the acceptable range for both personal and epidemiological purposes than to define an arbitrary scale of obesity.

While it is valuable to define the level of overweight in this way it is also important to take account of body image perception. This may depend on social, cultural and national differences. For example women in West Africa are considered more attractive if they are comely.

Hence when the practitioner discusses the target weight with a prospective dieter, the weight which is most likely to be achieved and maintained is usually better related to perceived body image rather than based upon actuarial considerations. Adherence to any diet depends upon the will power of the dieter, and this is likely to be greater if an acceptable target is established. However, that does not relieve the practitioner of the need to define the actuarial range and to encourage the patient to attempt to achieve it.

THE PREVALENCE OF OBESITY

Most of the studies have been undertaken in industrially developed countries and these show a somewhat higher prevalence of obesity in women than men.

One recent study[5] demonstrating the level of obesity in the United

Table 2 Percentages of obese people (20% above the top of the normal weight range) at age 36 according to social class and sex

Social class	Men	Women
I	4.2	7.1
II	4.8	3.4
III	5.2	6.0
IV	5.6	8.8
V	9.5	20.6

Based on Braddon et al.[5]

Kingdom is shown in Table 2. This indicates the situation relative to sex and social class in one age group. In women obesity is seen more commonly among the poorer classes, and this is true of most surveys in industrially developed countries. Different surveys show different prevalence in the different male socioeconomic groups. This particular survey also shows a rather minor increase among the poorer classes in men in the United Kingdom. Other recent studies in the United States[6] and the Netherlands[7], for example, also show that subjects of low socioeconomic status run the highest risk of obesity. The difference

is, in many cases, due to social factors. Among women who are well off, the present fashion cult is for a slim line. Hence obesity is avoided as far as possible. A similar fashion cult does not exist for men.

Taken broadly across the UK population we find that between 1 in 7 and 1 in 5 adults are at least 20% over the desirable weight, equivalent to around 8 million people. Another recent review of the UK situation suggests an even higher figure[8]. In the United States[9] and some other industrially developed countries, the proportion appears to be even higher.

PROBLEMS ASSOCIATED WITH OBESITY

For a considerable time it has been known that obesity leads to a series of problems: problems of social adjustment; increased incidence of certain specific disorders; increased morbidity with degenerative diseases; and above all a higher overall mortality.

MORTALITY

Most of the information relating to the effect of obesity on mortality was derived from the statistics published by the insurance companies, and the key paper by Lew[10] is still widely quoted. The ANIH Consensus Conference has also reviewed the available evidence[11].

The overall increase in mortality is approximately 15% for every 10% that the person is above the normal weight. Expressed in a slightly different way this means that each 10% above the normal weight reduces the life expectancy by at least 1 year. However, if an overweight person slims down to a normal weight longevity tends to be restored to the normal value.

The association between increased mortality and obesity is greatest in the group that is found to be overweight as young adults. The increase in mortality comes in the age group 40–60 years.

Hence all the available evidence suggests that obesity increases middle-age mortality, and that weight reduction is desirable to reduce the mortality.

Disorders specifically associated with obesity are as follows:

Hypertension

A number of well-controlled studies have demonstrated an association between hypertension and obesity. Of these perhaps the most valuable are those of Keys et al.[12] and the Framingham Study[13]. Very recently the whole subject has been reviewed by Staessen et al.[14]

Increasing degrees of obesity are associated with a rise in blood

pressure and the prevalence of hypertension at all ages increases with increased weight. This relationship stood up to a seven-nation, cross-cultural study[12], and 60% of those who lost weight experienced a reduction in blood pressure; many becoming normotensive.

Atherosclerosis and coronary heart disease

Obesity has therefore been shown to be associated with hypertension, and the same studies have indicated that obesity is also associated with hypercholesterolaemia.

Both obesity and hypercholesterolaemia are risk factors for coronary heart disease. If the two factors of hypertension and hyper-cholesterolaemia are excluded there is considerable dispute about whether there is a direct relationship between overweight *per se* and atherosclerosis. But this is largely academic in the light of the other associations.

However, when atherosclerotic heart disease is present, obesity increases the incidence of angina and the risk of sudden death. This may not be due to a direct effect on the disorder but to the increased oxygen demands of the obese individual and to the physical effects on lung movement, both of which produce an oxygen debt.

In contradistinction to the excess mortality in coronary disease, the Framingham study[13] could find no evidence of correlation between obesity and intermittent claudication or with cerebrovascular accidents (with the exception of that due to the associated hypertension).

Diabetes

It is clear that obese subjects develop diabetes mellitus more often than those of normal weight or the lean. Thus for example in the Framingham study, diabetes developed three times as often among obese individuals[13].

The aetiology is not yet clear, but it is known that, in the obese, circulating insulin is increased and there is some evidence of a peripheral resistance to the action of insulin. It is suggested that the increased production of insulin in the islets of Langerhans ultimately leads to a failure of the beta cells. This failure is coupled with continuing high needs from the obesity.

In those who develop diabetes, death from the disease is four times more common in those who are obese than in those who are of normal weight.

7

MORBIDITY

Gallstones

There is a direct association between obesity and the development of gallstones. Thus, for example, in one study[15] women below the age of 50 who had gallstones were 25% heavier on average than women without gallstones. The cause is probably an increased biliary excretion of cholesterol as a result of the hypercholesterolaemia of obesity.

Other disorders

Obesity is associated with an increased incidence of gout, hiatus hernia, hernias and varicose veins. Amenorrhoea and infertility are also common in the obese. Whether the incidence of some forms of cancer (e.g. endometrial carcinoma) is also increased in obesity is still a matter of dispute. Obesity can also precipitate skin disorders (as a result of various organisms multiplying in warm moist skin folds).

Apart from a definite correlation between the incidence of such disorders and obesity it is clear that excessive weight can increase the morbidity in many chronic diseases by imposing an additional strain on the body. These include chronic heart disorders, respiratory disorders, some gastrointestinal diseases, but above all arthritic disorders, particularly osteoarthritis of the weight-bearing joints. It also increases the difficulty in achieving mobility in those with chronic neurological diseases.

Another major problem of obesity is an increased level of risk during surgery. Not only are the technical problems of the surgeon increased, but the lack of general and chest mobility leads to an increase in respiratory infections, venous thrombosis and embolism in the postoperative period.

Social problems associated with obesity

The existence of an inappropriate body image can produce a large number of social and psychological problems, particularly in those who fail to come to terms emotionally with their increased size.

Among the numerous social problems that can be encountered are:

1. Fatness and gynaecomastia leading to embarrassment and teasing in adolescents, reducing participation in sports, increasing the sedentary way of life and hence further weight increase.
2. Failure to attract the opposite sex, or if the obesity occurs after marriage an increasing disenchantment leading to further social and sexual difficulties.
3. Loss of job opportunities, either through defined physical causes

(e.g. need for physical mobility) or through difficulties at interview.
4. Social isolation leading to depression.

Increased accident risk

In addition to the other difficulties experienced by those who are obese, there is clear evidence that obesity increases the risk of accidents, particularly those which occur in the home and workplace.

2

The physiopathology of obesity

THE SIMPLE EQUATION

In its simplest form obesity, consisting of the increase in the number of fat cells and the quantity of neutral fats that they contain, represents an intake of food in excess of the body energy needs (Figure 1).

The body takes in energy in the form of carbohydrates, fats, proteins and alcohol. All these substances can, by appropriate metabolic pathways, be converted into energy for body activity. This energy can be expressed in terms of kilocalories (1000 calories) or as kilojoules. One kilojoule is equivalent to 0.238 kcal. The different food components provide different levels of energy per gram utilized, viz.:

Carbohydrate	4 kcal
Fat	9 kcal
Protein	4 kcal
Alcohol	7 kcal

When there is a balance of food intake and energy needs the body utilizes the available calories from each of these sources for energy production. However certain tissues (e.g. the central nervous system) have pronounced preferences (i.e. in the case of the CNS for glucose). The interactions which ensure that adequate energy substrates are available at the required levels are largely controlled by hormones released from the endocrine glands.

The normal body also has reserves of energy. Carbohydrate stores, for example, are available in both liver and muscle as glycogen, though the amounts are small and only cover the needs of about 1–2 days of complete starvation. Fat is stored extensively both subcutaneously and around internal organs, while under conditions of total starvation the body may use proteins from the body tissues (mainly the muscles). When the intake of these energy-providing food substances exceeds the body needs then the majority of the additional material is converted (Figure 2) into fat which is deposited in fat cells throughout the body. When fat is stored, protein levels also increase. This is considered in detail later. The only circumstance in which excess energy can be excreted from the body rather than being stored is during gorging, when it leads to vomiting.

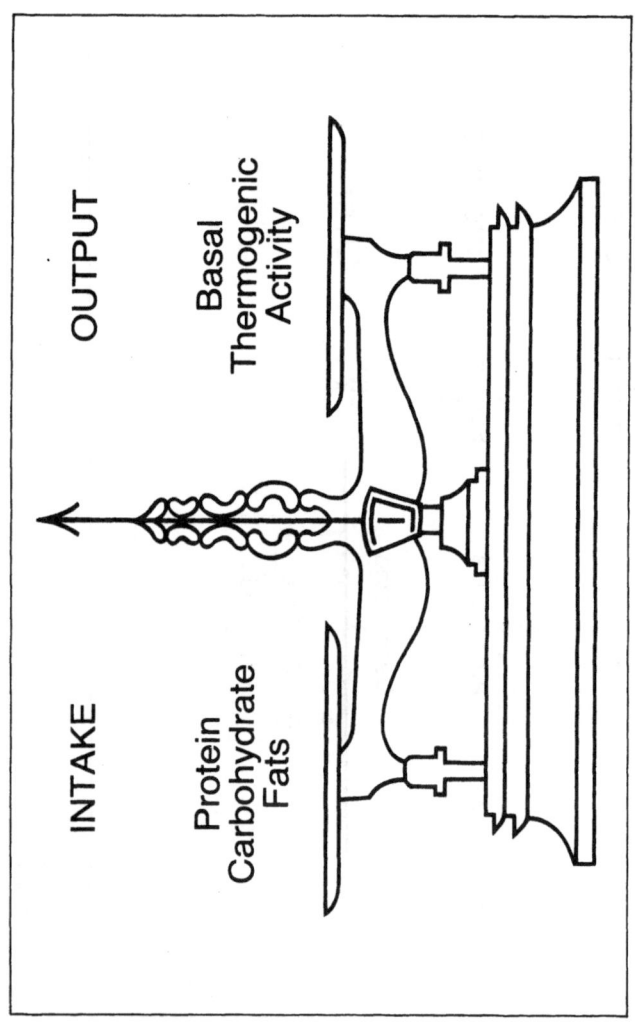

Figure 1   The balance between energy intake as food and energy output

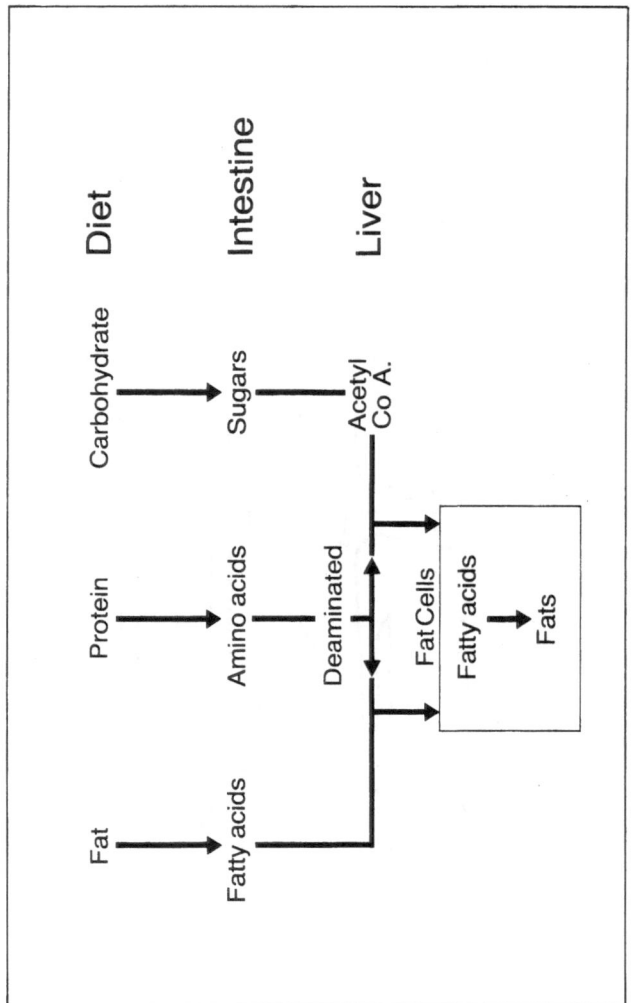

Figure 2 The pathways for the conversion of excess quantities of constituents of the food into fat tissue

The normal energy needs of the body consist of a basal level, representing mainly the maintenance of the resting body processes together with the body temperature; an additional energy need during the ingestion and metabolism of the food; and a variable energy output which depends on the level of physical activity.

The basal metabolic needs of the body vary with a number of factors which include the age, sex and size of the individual, and are also influenced by external factors of which perhaps the most important is the environmental temperature. Basal metabolic rate is higher in infancy and youth and declines with age (Figure 3). The basal metabolic rate is higher in men than women, partly because men have a higher lean body mass and partly because the higher level of subcutaneous fat in women reduces heat loss. The basal metabolic rate is also influenced by hormones, of which the most important are those of the thyroid gland (T_3 – tri-iodothyronine and T_4 – thyroxin) and the catecholamines.

The determination of the metabolic rate has been extensively improved over the past few years, and as these improvements have taken place it has become apparent that not only is there greater variation between normal individuals than was previously recognized, but the basal levels may be substantially lower than had been thought previously – particularly in some people. This applies whether the rate of metabolism is expressed in terms of the total output, or more reliably in comparative terms, expressed in terms of unit of weight or body surface area. The most accurate representation is to express it in terms of per kg of fat free mass. Fat free mass in non-obese adults normally represents about 70% of total weight.

The currently accepted range of normal levels is given in Table 3. As is explained elsewhere (page 17), while this represents our current view on the normal range, some entirely normal people as determined by clinical examination are metabolically very efficient, and have a substantially lower basal metabolic rate.

Metabolism is increased following food intake and with a normal mixed diet this leads to a metabolic rate rise of some 20% over the 24 hour period. This response is dependent to a certain extent on the nature of the food that is consumed, but for the average person extent of the variation with type of food is not sufficient to be a matter for concern.

Additional metabolic needs are created by physical activity and the increase depends on the amount and type of the physical activity. There have been few recent studies that have examined the additional energy requirements, and much of the older published information is based on methods that are now accepted as tending to lead to values which are too high. Adjusted figures, representing our current views[16]

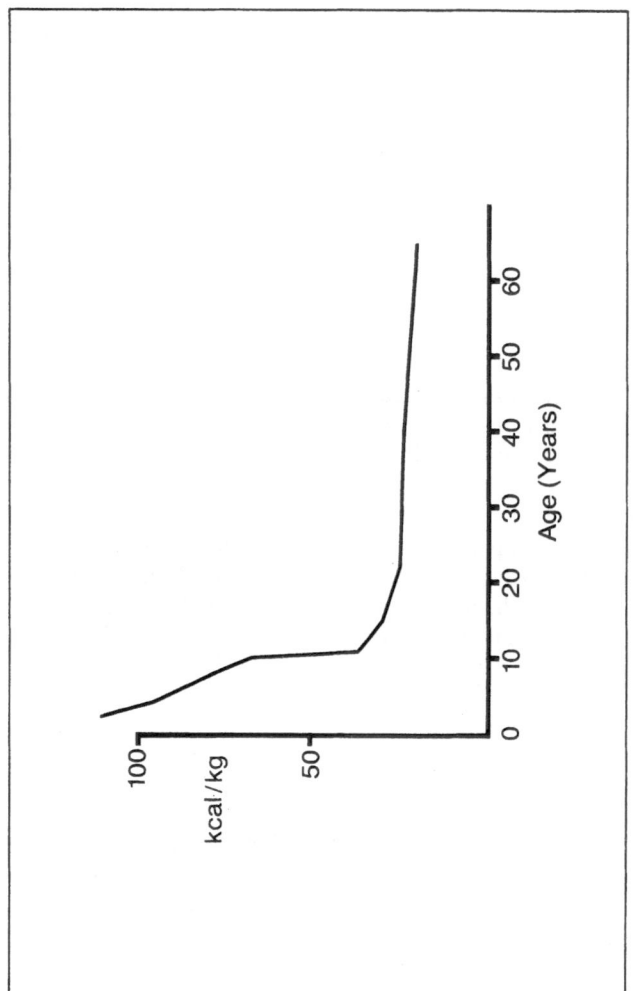

Figure 3 Variation of the basal metabolic rate with age. It is expressed in terms of the kcal output per kg body weight for males. Based on data in the WHO report[16]

Table 3 Basal metabolic rate (BMR) expressed as kcal per kg total weight and per day (rounded figures)

	BMR (kcal/kg)		BMR (kcal/day)	
Age (years)	Male	Female	Male	Female
1– 2		105		1150
3– 5		95		1550
7–10	78	67	2100	1800
10–12	36.5	33	1260	1190
14–16	29.5	26.5	1640	1380
18–30	25	23.5	1750	1300
30–60	24	24	1680	1320
60 +	20.5	21.5	1430	1180

For children average weights have been used. Adults are based on a 70 kg normal man and 55 kg normal woman. Derived from data on WHO monograph[16]

on the daily needs for various levels of activities are shown in Table 4. Samples of the factors for several physical activities[16] are given in Table 5. They should be regarded as a guide rather than as absolute figures, for apart from the limited nature of the data, the extent of the commitment of those involved in various activities will show substantial variation which will be reflected in the metabolic needs.

Hence the total calorie needs will vary markedly from one group to another, as will the percentage contribution of basal metabolism. For the majority of people leading a reasonably sedentary life, and this is the group which is most liable to suffer from obesity, the basal metabolic activity contributes a substantial proportion of the whole (about 70%).

Table 4 Average daily energy requirements of adults whose occupational work is classified as light, moderate, or heavy, expressed as a multiple of BMR

	Light	Moderate	Heavy
Men	1.55	1.78	2.10
Women	1.56	1.64	1.82

From WHO monograph[16]

THE CONVERSION OF ADDITIONAL CALORIES TO FAT

When the calorie content of the food exceeds the metabolic needs the additional food – whether it be in the form of protein, carbohydrate or fat – is converted into neutral fat and stored in adipose cells, either subcutaneously or around internal organs.

15

Table 5 Energy needs for various selected activities presented as a proportion of the basal metabolic rate (BMR)

	Male BMR ×	Female BMR ×
Standing quietly	1.4	1.5
Walking (normal pace)	3.2	3.4
Light cleaning	2.7	2.7
Cooking	1.8	1.8
Office work at desk	1.3	1.7
Light industry	2.0–3.6	2.5–3.4
Bricklaying	3.3	
Mining with pick and shovel	6.0	
Light recreation, e.g. bowls, golf	2.2–4.4	2.1–4.2
Strenuous recreation, e.g. athletics, rowing	6.6 +	6.6 +

Based on data in WHO Monograph[16]

The minimum number of additional calories that are required to lead to the deposition of 1 kg of additional fat is theoretically 9000 kcal. However since adipose tissue consists of fat, protein and water, a more accurate representation is 7000 kcal per additional kilogram of deposited fat tissue. Hence for a weight gain of say 1 kg per week, the daily additional calories above the needs would be approximately 1000 or an intake of approximately 3200 kcal in a person with a sedentary lifestyle.

COMPLICATING FACTORS IN THE EQUATION

Although the above principles apply as a basis for the understanding of the pathophysiology of obesity, several factors complicate the picture and make it more difficult to express the equation in absolute terms, viz.:

Thermogenic response to the intake of food

As has already been noted, food intake leads to an increase in the energy output. Hence if the food intake is increased above the normal requirements, a portion of the extra calories is dissipated in normal people by the thermogenic action. This probably applies also to a substantial proportion of those who become obese. However recent studies have indicated that a proportion of the obese show a reduced thermogenic response. This may be one reason for some patients becoming obese, although the proportion who fall into this category is probably small.

Efficient metabolism

Attention has already been drawn to the fact that a proportion of clinically normal people appear to have a greater level of metabolic efficiency than most people. Hence their basal metabolic rate is well below normal. This appears to be associated with a smaller than normal thermogenic response in some cases. Whether the metabolic needs for a given level of activity are also lower has not been determined.

This means that the balance of intake versus output is achieved at a much lower level of intake than in normals. Even a traditional reducing diet of say 1000 kcal may lead to an increase in weight in this unfortunate group of people.

Increase in lean body mass in the obese

A group from the Dunn Nutrition Laboratory in Cambridge[17] showed in 1978 that as people gain weight, their lean body mass also increases in proportion. They found that the lean body mass increase represented an average of 32% of the excess weight in women and 36% in men. It is not clear whether this represents an increase in the muscle tissue of the body or an increase in the cellular component of the adipose tissue or both. The latter appears to be the more likely explanation.

However, as a direct consequence of the increase in this lean body mass, the metabolic rate rises. That is to say, the obese have a higher basal metabolic rate. This will have the effect of reducing the amount of extra fat which is laid down as the weight increases – an attempt to self-correct.

As a corollary to this, as the weight falls during dieting, by whatever means, the relevant lean body mass is lost – unless at the same time sufficient strenuous exercise is undertaken to convert the released protein from the weight loss into muscle. As the lean body mass declines the metabolic rate automatically falls. This means that a deficit in the calorie intake produces a smaller weight loss the longer that dieting continues, whatever the method of dieting. This is another reason why some of the traditional diets fail to achieve a persistent weight reduction.

THE REASONS FOR OBESITY

A recent study[18] has demonstrated that a major factor in the genesis of obesity is the fact that a substantial proportion of those who are overweight underestimate their food intake.

In 13 lean people the estimated dietary intake corresponded accu-

rately to the calculated total energy utilization over the day. On the other hand in nine obese subjects the estimated dietary intake represented on average only 67% of the calculated total intake. That is to say the obese group were actually consuming on average some 877 kcal more than they estimated. Within this overall factor the available evidence suggests that there are four reasons why people become overweight for more than a brief period (e.g. Christmas, when most people overeat, underexercise and gain weight for at least a few days).

1. *Bad eating habits*. These people gain weight because their eating pattern does not correspond in any way to the accepted idea of a balanced diet. These are often called 'Junkaholics'. They eat too much food of poor nutritional value, consisting mainly of fat, sugar and alcohol. They frequent the fast-food shops and usually have at least one fried item (usually more) during each main meal. Between mealtimes they tend to have further fat- and carbohydrate-rich snacks. Their main problem is a lack of education about good nutrition, and nutrition education is difficult to achieve.
2. *The social eater*. This group of people become overweight through professional or social commitments. They either have to attend, or enjoy attending, innumerable social gatherings at which eating is part of the social requirement. These will include business lunches, dinners, cocktail parties and receptions. Such people often consume an excessive quantity of alcohol, and the good food provided for them tempts them to eat more than their energy needs require. Their willpower is often good – except when it comes to food and drink, both of which they enjoy. This group will include business executives, politicians and senior academics. It also includes celebrities and the jet-setting rich. The existence of slim members of each of these groups should remind us that it is not necessary to be a gourmand if one is a gourmet – it just requires willpower.
3. *The binger*. These people become obese because they take refuge in eating as a means of attempting to solve their social or emotional problems. Their response to a stress situation is to binge, using food rather as the infant uses the dummy. They are overweight for genuine psychological abnormalities, sometimes anxiety, but frequently depression. Their relationship to food is very much akin to that of many alcoholics to alcohol. They use it as an escape mechanism.
4. *The metabolically efficient*. A relatively small proportion of the obese gain weight because their metabolism is set very low; hence they need little energy for normal daily activities. Such people may need as little as 800 kcal or less a day for perfectly normal behaviour,

18

and if they consume more they will inevitably gain weight. They show none of the classical clinical signs of hypothyroidism despite their low metabolic rate. A proportion seem to have a hereditary basis, for they come from families in which one of the parents was overweight. In this case they usually give a history of being overweight from childhood. Some others report a sudden change in their weight following a serious illness, an operation or childbirth.

It is difficult to provide an accurate view of the proportion who fall into each of these categories; it probably differs from one practice to another. Certainly the experience in a specialist obesity clinic cannot be regarded as representative, for such a clinic gathers together a preponderance of those who have failed to lose weight adequately on other diets.

However, a clue to the main cause may be gained from the fact that the greatest proportion of overweight people is to be found in the lower socioeconomic classes. This suggests that an important cause in a substantial proportion of people is faulty nutrition, i.e. they fall into class 1 above. A further substantial group is that of the bingers, who use food to compensate for social and emotional problems.

3

Methods and problems of weight reduction

The aim of all weight reduction methods is to bring the patient's weight down to an accepted target weight which meets the requirements of the physician, in terms of the ideal weight, and the requirements of the patient, in terms of body perception and the ability to maintain the weight.

Hence any method of weight reduction should be judged not only in terms of the ease with which the target weight can be achieved without side-effects or adverse effects, but also the ability to maintain this weight after the target has been achieved.

Based upon the physiological considerations outlined in the previous chapter, in theory weight can be reduced by reducing food intake into the metabolic pool, or increasing the energy expenditure.

To date there are some 14 or so methods (in addition to very low calorie formula diets – e.g. the Cambridge Diet) that have been used sufficiently extensively for comments to be made about their value and problems. They can be divided into three broad categories, although there is some overlap of mode of action in relation to some of the methods. These categories are:

1. some form of non-invasive assisted voluntary food reduction,
2. methods of increasing energy output,
3. invasive surgical procedures.

It is not unusual for more than one method to be used simultaneously.

NON-INVASIVE REDUCTION IN THE FOOD INTAKE

Moderation in food consumption

For a relatively few overweight people moderation in food consumption will bring them down to a normal weight. This is particularly true if they are only just above the target weight. For many, however, moderation as a means of weight reduction is nothing more than a myth.

The reason that moderation will not work is that many people become overweight because they cannot achieve moderation in food intake. Hence the chance of reducing their weight by maintaining moderation is very small.

People who fail on diets often believe that this failure is due to lack of willpower or lack of motivation, or failure to estimate the calories adequately, or just cheating. This does not take account of the fact that as the weight falls the metabolic rate automatically also falls. Hence moderation becomes self-defeating because the person has to impose a greater level of moderation as time goes by, and this is difficult to achieve. Hence moderation is very poor for achieving a loss of weight in the first place.

Moderation is the ideal method for maintaining target weight once the target weight has been achieved, particularly if this is combined with sensible selection of foods.

Behaviour modification

The development of obesity is clearly the result of undesirable behaviour in the individual. This may stem from enjoyment of food, desire to relieve depression, undesirable social or work habits, etc. Ultimately, however, it can be regarded in behavioural terms. Hence the weight reduction and maintenance of an acceptable target weight can be regarded as behaviour modification.

The psychological approach of obese people to food appears to be quite different from those who are thin. The obese eat more quickly, take larger mouthfuls of food, chew their food less and drink far more with their food. In addition they do not necessarily eat to satisfy hunger but respond to food 'cues' such as the time of day and sitting and watching an advertisement on television. Based on these abnormalities, Richard Stuart[19] developed his method of behavioural therapy which attempted to change life-long eating habits.

His and other similar methods are both time-consuming and expensive, although they can be used in the form of group-orientated behavioural modification. This is less time-consuming and has been reported in the past to be more successful than conventional methods of weight reduction. In the main these reports have dealt with either institutionalized or middle-class groups. Investigators have addressed the need, and have published the results of preliminary studies of the long-term free-living behavioural modification programmes within the population. However the number of large published studies is small. The weight reduction achieved usually amounts to only a few pounds even in those who are grossly overweight.

These results are achieved at very high cost in physician and dietician time and cannot be regarded as good. Hence on the basis of the available evidence it appears that behavioural modification, however good it is in theory, does not produce satisfactory relief of obesity in practice.

Conventional diets between 800 and 1500 calories

The two types of conventional diets which are employed are either 'calorie counted' or 'carbohydrate restricted' diets. If they are composed of a wide variety of foodstuffs then the likelihood of vitamin or mineral deficiency is remote, provided that at least 1200 calories are consumed. In the range of 800 to 1200 calories it is very difficult to ensure that conventional diets provide vitamin and minerals intake at a level equivalent to the RDA.

Unfortunately conventional diets are not very effective. This would be anticipated from the pathophysiology involved. As the weight falls so the basal metabolic rate falls. Hence for many individuals the metabolic utilization often falls to a level close to that which is being provided in the diet.

The energy deficit is so small that the weight loss is unacceptably slow. When coupled with the bother of calorie counting it leads to the diet being abandoned.

Formula diets in the range of 1000 calories

With formula diets such as Metrecal (Mead Johnson) and Slender (Carnation Co.) the RDA of the most important minerals and vitamins are provided. Clinical experience with these two diets has been poor. The explanation for this is probably similar to that leading to the overall poor results with calorie counted or carbohydrate restricted diets between 800 and 1500 calories. Patients rarely adhere to these formula diets for more than 6 weeks, and if they do the weight loss is disappointing thereafter. Thus, although the diets are nutritionally good, the overall results are poor.

Very low calorie conventional diets

The use of protein-based diets is not new. Evans, Strang and McCluggage published a series of papers between 1929 and 1931 claiming that a low calorie diet of 400–600 kcal (1.68–2.52 MJ) could be used safely for the treatment of obesity. The final composition contained 400 kcal (1.68 MJ) and consisted of 23 g carbohydrate and 50 g protein[20,21]. Despite supplementation with vitamins and minerals the diet was still

deficient in a large number of nutrients.

Patients were maintained on the diet at home without complications for 6 months and longer. In 295 patients studied between 1929 and 1936 the average period of dieting was 8 weeks, the average loss 9.9 kg, or 1.2 kg/week.

More recently the Simeon[22] diet has been used extensively by many medical practitioners, in conjunction with injections of human chorionic gonadotropin (HCG). Although the latter were shown to have the same effect on weight loss as saline, patients were strongly motivated to continue for 5–6 weeks, during which time the weight loss was 9–14 kg. The diet contained about 600 kcal (2.52 MJ) and consisted of meat or fish, green vegetables, crispbread, and fruit. However, like any diet under about 1200 kcal based on conventional food, deficiency of minerals and vitamins is inevitable without supplementation. Over the relatively short period over which the diet was used these deficiencies would not be a serious consequence, but the diet is not recommended for longer than 6 weeks without appropriate supplementation.

Zero calorie diets

The so-called 'zero calorie diet', consisting of the intake of no energy-supplying material, achieved great popularity in the treatment of gross obesity a few years ago, and until recently was still being practised with enthusiasm in certain countries. Ideally the patient spent periods of several months in hospital, although shorter periods of up to 4 weeks have been effective in selected outpatients. The regime consisted of no food at all, but water ad libitum with vitamin and mineral supplementation.

The diet is extremely effective in terms of weight loss, but unfortunately serious complications can arise during treatment and starvation diets are not without considerable danger. Death due to ventricular fibrillation has been reported in several individual cases, including young apparently fit though overweight individuals. Some of the deaths have been associated with histological loss and fragmentation of the myofibrils of the myocardium.

In complete starvation the excretion of urinary nitrogen is high during the first week but it then levels out at about the fourth or fifth week to 3.6 g nitrogen per day in men and 4.2 g nitrogen per day in women[23]. Over a prolonged period signs of protein deficiency appear. Anaemia can develop and neutropenia and hypoproteinaemia have all been reported. Alopecia commonly occurs in both men and women.

Starvation causes an excessive loss of potassium, sodium, calcium, iron and other electrolytes. Carbohydrate has electrolyte-retaining properties. Lack of potassium causes muscular weakness, mental con-

fusion and electrocardiograph changes (S–T depression and broad low T waves). It is probably the potassium lack rather than the protein depletion which leads to sudden death. Sodium depletion can lead to hypotension and muscular cramps, calcium depletion to osteoporosis, and iron deficiency to anaemia.

Another consequence of the carbohydrate lack is an elevation of serum uric acid to pathological levels; this has led to attacks of gout in a few patients. Lactic acidosis leading to a fatality has been reported[24]. The chief pathological effect is a fatty liver.

Serum ketones have a pharmacological action not dissimilar to alcohol, and patients exhibit all the psychological symptoms of the alcohol drinker. The patient's mood and attitude is variable, mostly euphoric but occasionally aggressive or depressed. The euphoria seen in the first week of starvation is a large contributory factor in the success of 'health farms', since the client feels well, and also believes the excellent weight loss is chiefly body fat, whereas a large proportion of the first week's weight loss is fluid.

Starvation produces (as does any other weight reduction method) a rapid fall in basal metabolic rate of 20% or greater. This is principally due to a fall in serum T_3 since the conversion of T_4 to T_3 is by-passed and an inactive isomer – reverse T_3 – is the main metabolite[25]. In consequence a mild hypothyroidism develops, and the patient often complains of feeling cold.

Some of the hazards of starvation can be prevented by giving a multivitamin tablet containing, in addition, potassium (50 mEq), sodium (40 mEq) and iron (18 mg) per day. In addition, allopurinol can be prescribed to correct the hyperuricaemia. In the opinion of the authors zero calorie diets should only be used under strict medical control within hospital and with electrocardiographic and electrolyte monitoring to ensure that heart and mineral abnormalities do not occur. Whether the zero calorie diet is ever justified now is a matter of considerable dispute, though before the advent of very low calorie formula diets it was one of the very few effective methods of weight reduction.

Protein-sparing diets which led to 'liquid protein' diets

To achieve nitrogen balance in complete starvation, supplements of proteins such as egg albumin or casein can be given. Between 40 and 60 g protein per day is needed to obtain a positive nitrogen balance (Figure 4) with a carbohydrate-free diet[26]. The strongest advocates of the protein supplemented fast have been Blackburn et al.[27,28] who gave patients 100 g casein/day with some mineral supplements. Weight losses achieved were not dissimilar to those seen in complete starvation.

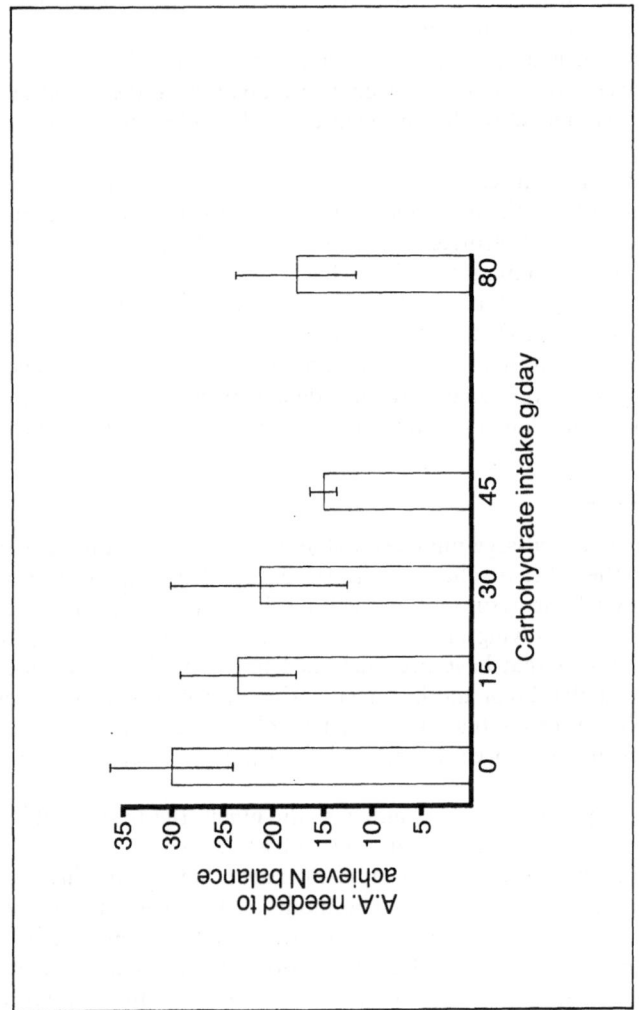

Figure 4 The effect of carbohydrate intake on amino acids (protein) required to achieve nitrogen balance (studies by Howard and McLean Baird)

The concept is good and effective. Unfortunately when it was developed commercially misinterpretation of the requirements led to major problems. The commercial preparations were based on hydrolysates made largely from cowhide, collagen and gelatin, to which saccharin and artificial flavouring were added. The proteins included in such diets were nutritionally of low biological quality, and did not contain an adequate balance of essential amino acids. Minerals and vitamins were not included. Not unexpectedly, several deaths ensued and, of these, ten were shown definitely to be linked to the 'liquid protein' diet[29].

It appears from the clinical and autopsy reports that the deaths may have resulted from the poor-quality protein (missing essential amino acids), from low electrolytes (probably mainly the potassium); low vitamins; or a combination of these effects.

It is important to appreciate that in current very low calorie diets (e.g. the Cambridge Diet) all these factors have been corrected. Hence these diets bear no similarity to the commercial 'liquid-protein' diets. Moreover extensive studies have not demonstrated any of the electrocardiographic changes encountered with the 'liquid-protein' diets.

Anorectic drugs

There have been a large number of studies of the use of anti-obesity drugs for the relief of obesity. The Food and Drug Administration have analysed the results of more than 200 short-term studies of various drugs involving almost 10 000 patients[30]. They concentrated on those studies that have been designed in the double-blind mode with either a placebo or another agent in the control group. The range of anti-obesity drugs that were used involved the majority of the substances that have been recommended for this purpose, as shown in Table 6.

In the early stages amphetamine was the main drug to be used but the dangers of dependence on this substance, coupled with the unpleasant side-effects and the rapid tolerance, have made this drug unacceptable for this purpose. The mode of action is on the catecholamine systems, and amphetamine has now been replaced by a number of different drugs with similar properties, viz. phenmetrazine, diethyl propion, phentermine and mazindol. Although the side-effects are less, they are still evident and include nervousness, irritability, insomnia as well as drowsiness, lethargy, dizziness, dry mouth, nausea, constipation and loss of libido.

Another drug commonly used is fenfluramine, which acts primarily on the serotonergic system. Side-effects are much less but a few complain of drowsiness, depression, lethargy or a sensation of unreality.

Table 6 The main anti-obesity drugs and the loss of weight which was achieved at 12 weeks in female patients with refractory obesity

Approved name	Proprietary names	Dose (mg/day)	Weight loss after 12 weeks (kg)
Diethylpropion	Apisate, Tenuate	75 mg	1.1
Mazindol	Teronac	2 mg	1.2
Phentermine	Duromine, Ionamin	30 mg	2.7
Amphetamine/dexamphetamine	*	12.5 mg	2.9
Fenfluramine	Ponderaz	120 mg	4.1

* No longer commercially available
Based on data from Munro, J. F. Clinical use of anti-obesity drugs. In Munro, J. F. (ed.) (1979) *The Treatment of Obesity*. (Lancaster: MTP Press), pp. 85–122

More recently several patients have been reported to develop irreversible pulmonary hypertension, which is of far greater concern[31].

The efficacy of the drugs is only modest. In the FDA study, after 4 weeks treatment, subjects taking the active drug lost 0.56 lb (0.25 kg) per week more than those on placebo.

In refractory patients on conventional diets the total weight loss in 12 weeks varies from 1.1 kg to 4 kg according to the drug used (Table 6). This is very poor compared with a very low calorie diet alone, in which a mean weight loss of about 15 kg can be obtained in 12 weeks (page 34).

From these studies there is no evidence to support the contention that appetite-suppressant drugs will help to establish normal eating habits. Indeed the evidence would suggest that it is easier to maintain a weight loss on diet alone than it is to maintain the weight loss when drug treatment has been used. The individual response to such drugs is very variable, perhaps partly reflecting differences in drug absorption and partly reflecting the various metabolic adaptive changes that can occur in obesity. The extent of tolerance as a factor in the variable effect is far from clear.

Hence in terms of long-term effectiveness and abundance of side-effects the current appetite-suppressant drugs have little to commend them for weight reduction and weight maintenance, but there may be some justification for their use to help patients over a short-term eating problem.

INCREASING ENERGY OUTPUT

Physical exercise

Theoretically it could be argued that physical exercise, which increases energy expenditure, could be used as a direct method of reducing

weight, but there are three fallacies to this argument. First the obese find it difficult to undertake physical exercise both because of the extra weight which they are carrying and also because many suffer from minor respiratory or cardiovascular problems, or from damage to the joints as a result of the obesity. Any of these factors will reduce their ability to undertake the necessary physical exercise.

The second reason for the failure is that the amount of *additional* energy that is expended, except under very strenuous exercise, is relatively small (Table 5, page 16). Since the obese cannot undertake extremes of physical activity the energy deficit produced by the exercise which they are capable of undertaking provides a relatively small component towards their total daily activity.

The third reason for the failure of exercise to produce a satisfactory weight loss is that a high proportion of patients, even those who are most highly motivated, find it difficult to maintain a training programme for more than a rather short period. Hence compliance is rather poor, and it is estimated that even with maximal effort to keep obese subjects in a training programme, the drop-out frequency is about 30%[31,32].

Although exercise as the sole means of losing weight has a very limited effect, this is not to suggest that those who are undertaking a weight loss programme by other methods, and particularly the Cambridge Diet, should not at the same time try to get themselves more fit by a regular training programme. This will have the effect of improving the cardiovascular and respiratory activity and increasing the muscle tone. It will also increase both the total metabolic consumption and the basal metabolic rate (via an increase in the lean body mass) and hence will increase the rate at which the weight is lost.

Strenuous exercise should not be undertaken during the first few days, particularly in those who have been sedentary previously, or who are unfit.

Thyroid hormones

There have been several studies which have looked at the use of thyroid hormones, in particular T_3 (L-tri-iodothyronine) in the treatment of obesity. Most used pharmacological doses, with or without dietary restriction. One notable feature of all these studies has been the clinical tolerance to large doses of the thyroid hormones by obese patients.

Thus, for example in one study of the effects of either T_3 (225 μg per day) or placebo in conjunction with an 800 calorie diet[33] by the eighth week there was a significantly greater weight loss in the T_3-treated group but by the twelfth week the differences were no longer significant. The doses of the thyroid hormones used in these patients

would undoubtedly have rendered them biochemically, if not clinically, thyrotoxic. A potential hazard of this treatment in the long term is the development of complications such as osteoporosis and cardiomyopathy.

Hence the results of the use of thyroid hormones in the treatment of obesity are not outstandingly good, and there are potential risks in their use.

The use of T_3 as a short-term adjunct in some people taking the Cambridge Diet is considered on page 59.

SURGICAL INTERVENTION

Dental splintage

One method which is still in reasonably common use for patients who fail to maintain a regular pattern of small meal eating is dental splintage. The selection of patients for dental splintage is very similar to that for other surgical methods for the management of obesity. It includes long-term obesity which has failed to respond to other methods of weight reduction, excessive obesity and a reasonable motivation to achieve a weight loss.

The procedure itself is very unpleasant for the patient, although it is safer than other surgical procedures. On the whole very few patients will tolerate the dental splintage for a long enough time to enable a target weight to be achieved. Apart from the results of a poor nutritional status from the use of an inappropriate fluid diet (usually established at about 800 calories per day) other side-effects tend to be minimal. If the patient will tolerate the procedure then the weight loss is usually good.

Surgical procedures on the stomach

There are two main surgical procedures that have been carried out on the stomach for the management of obesity. The first is a gastric bypass in which the upper portion of the stomach is joined to the proximal duodenum. The operation has a relatively high surgical risk and hence should only be undertaken where all other methods have failed[34].

The weight loss was in fact good. However morbidity after the gastric bypass operation tends to be quite extensive, particularly the dumping syndrome which occurs in about one-third of the patients and a stomal ulcer which occurs in about 1.6% of patients and is a particularly troublesome one to treat. Late metabolic effects, for example osteomalacia, are relatively uncommon.

The second gastric operation is gastroplasty, in which the size of

the stomach is maintained but the passage between the proximal portion of the stomach and the distal is reduced. Long-term results are usually good and this is probably the surgical procedure of choice for intractable severe obesity which does not respond to non-invasive measures.

Jejuno-ileal bypass

Jejuno-ileal bypass enables weight reduction to take place as a result of iatrogenic malabsorption[35]. The technique was originally described some 30 years ago and was used to a fairly considerable extent in the 1960s. It is now much less commonly used. The procedure is a severe one, and in consequence of its radical and aggressive nature, the mortality of the operation and the side-effects, it should be considered only in those who have failed to respond to all other forms of therapy.

Within this framework the actual results on obesity were good. During the first year the weight was normally reduced by about 60–70% and although weight loss slows after this stage, most of the patients maintain a weight loss of around 35% of the original. It is excessively rare for the patient to experience less than a 20% weight loss. The main loss of weight comes from the fat tissue, although there is a reduction in the lean body mass in excess of that calculated from the loss of 'fat' tissue during the first 3 months.

The mortality is probably of the order of about 3–5%, mainly occurring during the first 2–4 weeks after the operation. There is considerable morbidity and among the main late problems may be mentioned diarrhoea, vomiting, fatty liver or cirrhosis, hair loss, deficiency of vitamins including folate and vitamin B_{12}, and abdominal bloating and wind.

Overall the procedure is only justified if all other methods have failed, and at the present time few surgeons would recommend undertaking this operation, favouring gastroplasty which has a lower mortality and morbidity.

Surgical removal of excess fat

The removal of excess fat, particularly from the abdomen (apronectomy), from the breasts, arms and thighs has been practised by cosmetic surgeons for some considerable time. The procedure may have some limited value for cosmetic purposes, but its value for the management of the underlying problem of obesity is extremely limited. It is only possible to remove a very small percentage of the excess weight, and unless the underlying cause is removed this weight is rapidly replaced.

Apronectomy may in fact be physically beneficial. By removing the

apron of fat within the abdominal cavity the centre of gravity may be allowed to return to a more reasonable position, which will allow a better posture, reduce strain on the weight-bearing joints, and allow a greater measure of physical exercise than had been possible in the past. However, surgical removal of fat has, apart from this, only a negligible role to play in the treatment of obesity.

4

The development of the Cambridge Diet

The development of the Cambridge Diet stems from the early 1960s when one of us (ANH) found that at the age of 30 he was not only becoming fat and unhealthy but, like so many others, could not easily take off the extra weight. Five to ten pounds (2–4.5 kg) weight could be lost easily by judicious feeding habits but minimal relaxation rapidly led to a return to the original weight again. He therefore decided to investigate methods of weight reduction using himself as one of the guinea pigs. This coincided with a period of interest in dieting procedures such that background information on methods and results was readily available.

Among these methods which had been employed was the technique of complete starvation (page 23). The results in terms of weight loss were dramatic. Unfortunately at least five publications recorded deaths in a number of patients treated by complete starvation. These deaths were due to severe degeneration of the heart muscle. Hence starvation could not be contemplated for long-term weight reduction because of the high risks, despite the impressive weight losses.

At this stage (about 1970) the standard teaching was that safe metabolic integrity could not be achieved with a diet which contained under 800 calories, though in the 1930s some studies, forgotten in the meantime, had been undertaken on diets of 400–500 calories. The research project, which was conducted in association with Dr Ian McLean Baird, physician at the West Middlesex Hospital, was to determine whether a formula diet could be devised which would have the excellent weight loss properties of complete starvation, but would be free from side-effects. The factors to be studied were:

1. The prevention of the cardiac muscle damage which resulted from complete starvation. It appeared likely that these effects on the cardiac muscle probably resulted from protein losses or from a combination of protein losses and electrolyte imbalance.
2. The avoidance of the severe ketosis which normally occurs during

starvation.
3. The avoidance of the effects of low carbohydrate intake on renal function.
4. The provision of adequate levels of each of the vitamins, minerals, essential fatty acids and trace elements to maintain health.

Experimental studies commenced at the West Middlesex Hospital in June 1970 on five severely obese in-patients. These five patients were all refractory cases, having failed on previous dietary regimes.

The initial studies used a mixture of the daily recommended levels of vitamins, minerals and a source of essential fatty acids to which are added various levels of protein and carbohydrate. Each study segment, designed to determine the lowest required level of protein, lasted 2 weeks, and the artificial protein used initially was a mixture of amino acids compounded for American astronauts. These initial studies omitted all carbohydrate from the diet and provided the total calorie intake in the form of the synthetic protein.

With protein alone, a nitrogen equilibrium was established after about 6 weeks with a protein intake of about 30 g/day (Figure 4). The significance of this in relation to the protein levels in the vital body tissues (internal organs and muscles) is considered later.

Small quantities of carbohydrate were then added. The addition of carbohydrate reduced the protein intake to achieve nitrogen balance (Figure 4) and at a carbohydrate intake of about 45 g/day the protein requirement was reduced by a half to about 15 g/day[36].

As a result of these trials it was found that the minimum requirements for the daily intake were protein 15 g and carbohydrate 30–45 g/day, together with the minerals, vitamins, trace elements and essential fatty acids, a total of between 200 and 300 calories per day.

Under this regime the five patients shed on average more than 2 kg per week, yet they felt well, experienced only minimal hunger and could be quite active around the ward.

A second trial was commenced in 1973 on 50 patients[37]. Due to the limited availability of hospital beds they were hospitalized for the first few weeks but then followed up as outpatients. Since the prepared amino acid mixture was too expensive for consideration for routine use, egg albumin was substituted as the protein source, since it is reputed to contain the ideal balance of amino acids and particularly the essential amino acids. The vitamin and mineral supplement was provided as a premix added to the diet itself.

Seventy-seven per cent were able to undergo a total treatment of 8 weeks, and 54% 16 weeks or over. The weight loss was substantial, 17 patients came within 15 kg of their ideal body weight and the average weight loss was about 1 kg per week. However, quite a lot of

the patients either found it difficult to remain faithful to the diet or even dropped out completely due in the main to the extremely unpalatable nature of the mixture. However over the whole 3-year period during which this second trial was undertaken, no important side-effects or toxicity were encountered. The patients felt and looked well.

It was considered that a combination of the diet together with an appetite-suppressant might be valuable. A double-blind crossover controlled trial of 20 outpatients in each group (i.e. with and without mazindol as appetite-suppressant) demonstrated the somewhat surprising result that patients could overcome the initial feeling of hunger as well without the appetite suppressant as with it[38]. The weight loss in each group was the same but, more important, the diet could be tolerated in outpatients without an unacceptable level of hunger. The 8-week study confirmed once again the safety of the product.

In the meantime (1976) the unpalatable egg albumin was substituted by dried skimmed milk as the source of protein and studies by food technologists were undertaken to produce a more palatable formulation with various flavours. These resulted in the definition of a complete diet comprising 33 g of protein, 44 g of carbohydrate and 3 g of fat with vitamins, minerals and trace elements equivalent to the daily recommended levels established by the Food and Nutrition Board of the National Research Council in the USA. This formulation has remained the basic formula ever since, apart from minor adaptations, mainly to achieve better palatability. This is the formula marketed as the Cambridge Diet.

The effectiveness and safety of this new formulation was tested not only within hospital but also in outpatients. Altogether a total of 50 obese patients were given the new formulation for from 4 to 12 weeks[39]. After 4 weeks the mean weight loss was 9 kg in hospitalized patients and 7.25 kg in outpatients. After 8 weeks *both* groups had lost a mean of 11 kg. This somewhat surprising result stemmed from the fact that at the sixth week the hospitalized patients had been discharged on an 800 calorie diet.

This study was perhaps the key one which led directly to the decision to make the diet generally available. Not only did it finally demonstrate that equally good results could be achieved in outpatients, but nitrogen balance studies in the inpatients demonstrated only small nitrogen losses, serum electrolyte levels were unchanged and electrocardiograms revealed no abnormalities associated with cardiac muscle damage. A summary of the comparisons of the effects of starvation and the Cambridge Diet is given in Table 7.

Subsequently in addition to further metabolic studies under the care of Dr McLean Baird at the West Middlesex Hospital, a University

Table 7   Comparison of starvation and very low calorie diets

Parameter	Starvation	Very low calorie diets (180–360 kcal)
Weight loss	Very good	Very good
Patient acceptance	Poor	Good
Ketosis	Excessive	Minimal
Serum uric acid	High	Normal
Nitrogen balance	Negative	Equilibrium achieved
Water balance	Diuresis	Normal
Electrolyte excretion (K^+, Na^+, Ca^{2+})	High negative balance	Equilibrium achieved
Safety	High risk	Theoretically safe

Department of Medicine outpatient obesity clinic was established in Addenbrookes Hospital. This provided extensive outpatient experience of the treatment of varying grades of obesity on patients referred by other hospital consultants and practitioners. Up to 30 or 40 patients were seen in this clinic per week, initially with laboratory control, but later as greater experience demonstrated the safety of the diet, under clinical control. This university department outpatient clinic still continues, and forms a useful centre to which patients in need of additional care can be referred.

This clinic at Addenbrookes Hospital has provided facilities for the study of several hundred obese people, many of whom have been referred there because they have not succeeded in losing weight with the normal dietary regimes provided by hospital dieticians. It has provided a spectrum of patients, including those who are relatively little overweight to those with a massive obesity problem, who have maintained the Cambridge Diet as their sole source of nutrition for periods of up to a year. Hence the long-term safety could be assessed – at least in a few people who were grossly obese and who needed these very prolonged dietary restrictions.

But in addition to these studies at the West Middlesex Hospital in London and at Cambridge, the diet has now been studied extensively by physicians in many other centres in Great Britain, Europe and the USA (Table 8). Some of these studies have been concerned with the important aspect of extensive clinical experience of the dietary use. Others have taken the form of more intensively investigated groups of patients. These have studied, *inter alia*, the effects of the Diet on the electrocardiogram, electrolytes, blood pressure and plasma lipid levels. Others have been concerned with use in patients with other disorders, including particularly diabetes.

Table 8 Centres in which clinical studies of the Cambridge Diet have been undertaken – to August 1986

United Kingdom	Cambridge	Sweden	Gothenburg
	London	Holland	Rotterdam
	Manchester	Italy	Naples
	Glasgow	Denmark	Copenhagen
	Edinburgh	DDR	Dresden
Ireland	Dublin	Poland	Stettin
USA	Atlanta	USSR	Leningrad
	Charlottesville	Finland	Helsinki
	Denver	South Africa	Johannesburg
	Rochester		Durban
	Newhaven	Oman	Muscat
	Los Angeles	Australia	Melbourne
	San Diego	Hong Kong	
Argentina	Buenos Aires		

Currently almost 50 clinical and scientific papers specifically related to the Cambridge Diet have been published (Appendix 1). This is in addition to books, popular and medical, about the Diet (Appendix 2) and to papers which have been concerned with very low calorie diets in general (Appendix 3).

Hence the published information on the effectiveness and safety of well-formulated very low calorie diets (e.g. the Cambridge Diet) is now extensive.

Part II

THE CAMBRIDGE DIET

5

The composition of the Cambridge Diet and how to use it for weight reduction

COMPOSITION

Sachets

The Cambridge Diet original formula is a compounded food in powdered form which, when added to water, provides soups or sweet drinks. The recommended intake of the powder (103 g per day) provides the US recommended dietary allowance (RDA) for vitamins, minerals and trace elements. The United Kingdom RDA is less extensive than the American counterpart but the formulation covers all the UK RDA needs. All these are provided within a daily intake of 330 kcals (Figure 5).

The amino acid composition is shown in Table 9, which also compares the daily intake on the Cambridge Diet with the USA RDA. The primary sources of protein are low-fat milk solids and soya flour, which provide at least 100% of the daily requirements of the essential amino acids together with most of the non-essential amino acids (Table 9). Fat is provided by the soya flour, which contains linoleic acid with lesser amounts of linolenic acid. Lactose is the primary carbohydrate. A dietary bulking agent is added.

The standard composition for the standard powder for vitamins and minerals is given in Table 10.

The only differences in formulation between the various forms of the final product, which is supplied in sachets, lie in the flavouring and colouring agents. These are all EEC approved for food use and the majority are natural substance.

Each sachet contains 110 calories, equivalent to one meal on the Diet, in a form which allows for ease in transport and use.

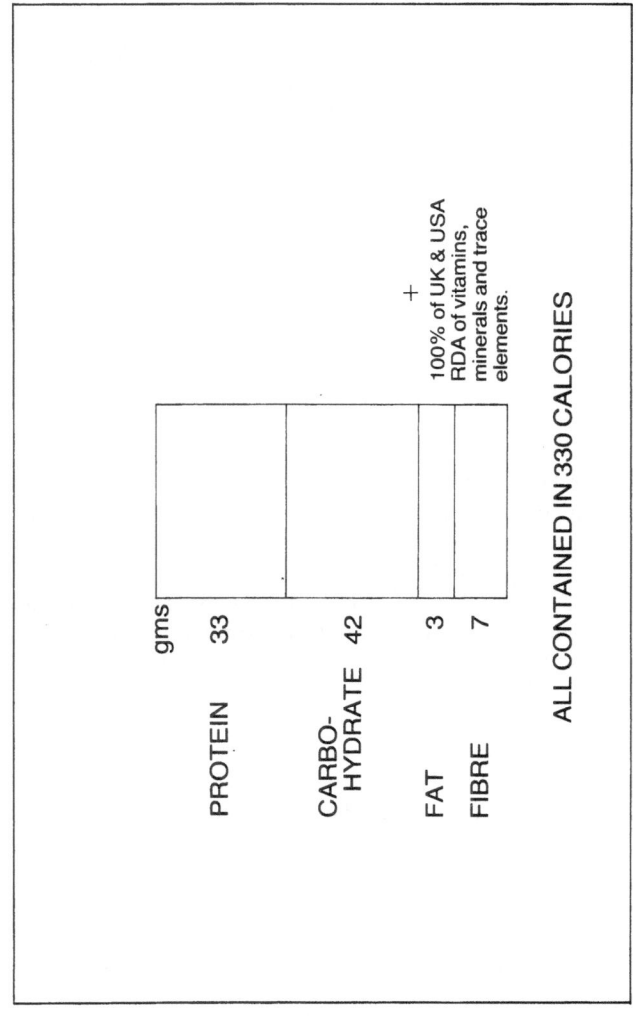

gms

PROTEIN   33

CARBO-
HYDRATE   42

FAT   3

FIBRE   7

ALL CONTAINED IN 330 CALORIES

+

100% of UK & USA
RDA of vitamins,
minerals and trace
elements.

Figure 5   The main constituents in the Cambridge Diet

Table 9 The Cambridge Diet – amino acid composition. Shown as per g protein in the diet, as intake per day on three Cambridge Diet meals compared with the USA RDA for the essential amino acids

Amino acid composition (g protein-bound amino acid per 100 g protein)		Per day (g)	USA RDA (g)
Aspartic acid	7.4	2.5	—
* Threonine	3.6	1.2	0.6
Serine ·	4.3	1.4	—
Glutamic acid	18.9	6.3	—
Proline	8.3	2.8	—
Glycine	2.0	0.7	—
Alanine	2.7	0.9	—
* Valine	6.0	2.0	1.0
* Methionine	1.5	0.5	0.7
* Isoleucine	4.4	1.5	0.8
* Leucine	8.2	2.7	1.2
Tyrosine	4.2	1.4	}1.2
* Phenylalanine	4.5	1.5	
* Tryptophan	1.3	0.4	0.2
* Lysine	6.4	2.1	0.8
Histidine	2.6	0.9	—
Arginine	3.8	1.3	—

* Essential amino acids

Nutrition bar

In addition to the sachets a new formulation has been introduced – nutrition bars in five different flavours. These nutrition bars have been developed to replace one serving per day of the Cambridge Diet. Designed on the same lines as the diet, the meal bar contains a balanced formulation of protein, carbohydrate and fat together with one-third of the US RDA for vitamins and minerals for each bar. Each bar contains 11 g of unchangeable carbohydrate per 58 g bar.

The meal bars are covered in chocolate, have a pleasant chewy texture and contain 140 calories (as opposed to the 110 calories in each sachet). The full range of available flavours of the sachets and meal bars is given in Table 11.

Making up the Diet

One sachet of Diet is added to 8–9 fluid ounces of water and shaken, blended or whisked either mechanically or with a fork. Either hot or cold water may be used, but not boiling water. Nor should the mixed diet be brought to the boil, since this reduces the vitamin C content.

Table 10 The Cambridge Diet – composition in terms of vitamins and minerals. Shown as per 100 g, per serving and per day of three Cambridge Diet meals compared with the UK RDA.

	Per 100g	Per serving (34.3 g)	Per three servings (103 g)	Percentage UK RDA***
Vitamin A	1.0 mg	0.3 mg	1.0 mg	133
Vitamin B$_1$	1.5 mg	0.5 mg	1.5 mg	125
Vitamin B$_2$	1.7 mg	0.6 mg	1.7 mg	106
Niacin	18.5 mg	6.3 mg	19.0 mg	106
Vitamin B$_6$	2.1 mg	0.7 mg	2.2 mg	*
Pantothenic Acid	6.8 mg	2.3 mg	7.0 mg	*
Biotin	194.2 μg	66.7 μg	200.0 μg	*
Folic Acid	388.3 μg	133.3 μg	400.0 μg	133
Vitamin B$_{12}$	2.9 μg	1.0 μg	3.0 μg	150
Vitamin C	58.2 mg	20.0 mg	60.0 mg	200
Vitamin D$_3$	9.7 μg	3.3 μg	10.0 μg	100
Vitamin E	48.5 mg	16.7 mg	50.0 mg	**
Vitamin K	135.9 μg	46.7 μg	140.0 μg	*
Calcium	776.7 mg	266.7 mg	800.0 mg	160
Phosphorus	776.7 mg	266.7 mg	800.0 mg	*
Magnesium	388.4 mg	133.3 mg	400.0 mg	*
Potassium	1.9 g	0.7 g	2.0 g	*
Sodium	1.4 g	0.5 g	1.5 g	*
Chloride	1.8 g	0.6 g	1.8 g	*
Iron	17.5 mg	6.0 mg	18.0 mg	150
Zinc	14.6 μg	5.0 mg	15.0 mg	*
Iodine	145.6 μg	50.0 μg	150.0 μg	107
Copper	2.9 mg	1.0 mg	3.0 mg	*
Manganese	3.9 mg	1.3 mg	4.0 mg	*
Selenium	58.3 μg	20.0 μg	60.0 μg	*
Molybdenum	145.6 μg	50.0 μg	150.0 μg	*
Chromium	58.2 μg	20.0 μg	60.0 μg	*

Figures rounded to one decimal place
* No UK RDA (recommended daily allowance) has been established, but the quantities stated are within the range of the adequate daily dietary intake specified by the National Academy of Sciences, USA or the US RDA
** 5 × US RDA (also used as antioxidant)
*** The equivalent figures for the USA RDA would be 100% with the exception of vitamin E which is also incorporated as a natural antioxidant

The Cambridge Diet is produced in several flavours (see Table 11) and this allows variation in flavour either by selection of a different flavour for various meals or alternatively using half a sachet made into 5 fluid ounces of hot soup followed by half a sachet of a sweet flavour.

Various other flavours can be achieved by the addition of no-calorie (or very low-calorie) flavours to either savoury or sweet basic formulae. In any such compounding it is important to ensure that a minimum number of calories is added and that no additional sodium chloride is

Table 11   Flavours in which the Cambridge Diet formulae are available

	Sachets		
Soups		Sweets	Meal bars
Minestrone		Vanilla	Malt toffee
Chicken		Strawberry	Caramel
Beef		Chocolate	Peanut
Asparagus		Banana	Chocolate
Turkey and vegetable		Peach	Lemon and ginger
Mushroom			

incorporated.

Under normal circumstances one serving is taken three times each day at meal times in place of any food as the sole source of nutrition. However, the following possibilities exist to deal with specific situations:

1. Six half-servings are taken at intervals of about 3 hours, particularly during the early stages to avoid hypoglycaemic reactions.
2. An extra full serving is taken each day particularly by large, active men. The extent to which active men feel that they need the additional quantity varies greatly, but for those who do feel the need an extra sachet of Diet is nutritionally better than a carbohydrate meal, for it maintains the mild ketosis which is so important for imparting the sense of well-being and avoiding hunger feelings.
3. Some people find great difficulty in drinking black coffee or tea (see below). For these dieters we have found it useful if they keep aside about a quarter-sachet of the vanilla flavour per day, diluting this to milk consistency and using it as necessary.
4. During the early stage of dieting using the Cambridge Diet as the sole source of nutrition we believe that there are advantages in only using the powder formula. On the other hand, one meal bar per day has advantages for those who have difficulty over facilities for mixing all their meals. It can be used as a substitute for one sachet.
5. While we favour the use of sole source Cambridge Diet for weight reducing purposes it can be used supplemented by one 400 calorie meal. A range of Cambridge packaged meals is available which may be used for this purpose (see page 82). The rate at which weight is lost is clearly less, hunger is more of a problem and adherence to the dieting programme tends to be less good (perhaps because of the lesser commitment which led to the desire for the meal). Nevertheless we have encountered several dieters who have been successful on the modified regime.

Other action that must be taken

In addition to the 24–27 fluid ounces of fluid that are used to dilute the three meals *at least another* 60–80 fluid ounces (3–4 pints) of fluid should be drunk each day. This is the most important advice for any dieter, particularly during the first week. The mild side-effects and discomforts experienced by some during the first week are nearly all explicable on the basis of an intake of fluid inadequate to compensate for the diuresis induced by the Diet, and hence to dehydration.

The three to four pints of fluid may be taken in the following forms:

Plain water

Weak black tea or coffee. China or herbal teas or decaffeinated coffee are preferable. Calorie-free sweeteners may be used. It is useful to enquire whether the dieter is a very heavy caffeine drinker before starting on the Diet, and if so to advise gradual reduction in the caffeine intake. We have encountered an occasional pronounced caffeine withdrawal syndrome when a dieter has taken the advice to abstain from caffeine too literally.

Low calorie fruit squashes well diluted.

Small quantities of 1-calorie carbonated drinks. These are particularly valuable when dieters have to attend parties or bars.

We would offer specific warnings about the use of the fluids shown in Table 12, which from bitter experience we know can interfere with the weight loss programme in dieters who were quite convinced that they were taking the Diet as sole nutritional source.

USE OF THE CAMBRIDGE DIET AS SOLE NUTRITIONAL SOURCE WHEN NOT UNDER MEDICAL SUPERVISION

Whether it is desirable or not, the majority of overweight people diet without being under medical direction or even without seeing their own practitioner. Cambridge Nutrition Limited, the Norwich-based company which markets the Cambridge Diet, have direct practical experience of this and they cannot foresee being able to change this immediately.

The Cambridge Diet, on the basis of over 10 years research, has been found to be very free from adverse effects in those who are overweight but otherwise normal. Accordingly the Company policy has been to *strongly advise* potential dieters to see their own doctors before they start the diet but not to *refuse* to supply the diet unless the potential dieter either suffers from a listed disease or is receiving medication. Under those circumstances, supplies are withheld unless there is agreement in writing to use them, provided by the practitioner.

Table 12 Fluids which should be avoided or seriously restricted when taking the Cambridge Diet as strict nutritional sole source

Avoided	Restricted
Those containing alcohol	Skimmed milk
Spirits	Strong coffee
Fortified wines	Low-calorie carbonated minerals, colas, etc.
Wines	
Beers	
Ciders	
Full-cream milk	
Fresh fruit juices	
Full-calorie squashes	
Full-calorie minerals, colas, etc.	

This is considered in more detail in Chapter 6.

A recent survey[40] has indicated that about 13% of potential dieters fall into the category where the practitioner's signature is required because of illness or medication. Of the remainder, 30% actually consult their doctor about dieting before they commence the use of the Cambridge Diet.

This policy does, however, mean that, like all dieting processes, a substantial proportion of those who are trying to lose weight using the Cambridge Diet are not under direct medical control. The absence of adverse reactions appeared to justify this policy. Extensive experience, both in the USA and the UK, has confirmed that problems do not arise as a consequence.

When the dieter is under medical control it is for the practitioner to decide how dieting should proceed. This is considered in Chapter 6. Dieters who are not under medical control are advised to take an extra meal of 400 cal per day for 1 week after 4 weeks on the Diet as their sole source of nutrition. There are no medical reasons for this, but the American authorities advised that this should be adopted when the Diet was first marketed, and that suggestion has been accepted.

After 1 week on an extra 400 cal per day the dieter can recommence the Diet as the sole source of nutrition in 4 week periods until the target weight is achieved.

RESPONSE TO THE DIET

During the first 24 hours on the Diet the glycogen stores are depleted, leading to a release of about three times the weight of water normally associated with the glycogen stores. The content of the alimentary tract is either absorbed and used, or alternatively is voided; in addition there is a marked diuresis. All these factors lead to a net fluid loss from

the body – even with the advised high water intake. In consequence almost everybody experiences an initial weight loss (almost entirely water, in fact) which encourages maintenance of the diet.

The next day or so may be uncomfortable for the dieter. At this stage several find that hunger is present (the ketosis has not yet occurred) and further weight loss is often small over the next few days (note that for a loss of 1 lb (0.4 kg) of fat 7000 kcals excess energy expenditure over intake is necessary, yet at this stage the metabolic rate has already fallen from its previous high levels during overeating by removing the dynamic action for the food). Also it is in this phase that the fluid intake often drops so that the side-effects may appear. The third day of the diet is usually regarded as the worst day.

Once through this stage the typical reaction is for a loss of weight which *averages* about 3–4 lbs (1.4–1.8 kg) a week, although this varies markedly from one person to another and from one week to another in the same person (see plateau – below). In the studies we have encountered weight losses of only about 1.0 kg per week in those who we felt were adhering to the Diet.

As the target weight is approached it is usual for the rate at which the fall in weight takes place to become slower. This is probably in part a natural phenomenon due to the metabolic rate: in part due to the fact that many people supplement their food intake to a greater or lesser extent at this stage, due to the boredom of sustained dieting.

OTHER BODY FUNCTIONS WHICH SHOW A POSITIVE RESPONSE TO THE DIET

During the extensive clinical trials that were undertaken prior to the introduction of the diet it became apparent that, in addition to the weight loss, there were important subsidiary beneficial effects of the Diet. These have been confirmed subsequently by other investigations.

Blood lipids

Use of the Cambridge Diet as sole nutritional source reduced the blood cholesterol by an average of 25% and the triglycerides by an average of 40%[39, 41–44]. All the patients in the study showed some reduction, and the decrease was greatest in those with initial high levels. The reduction only occurred while patients were using the Diet as sole source (Table 13), and when they returned to a normal diet the levels rose again, but not to pretreatment levels. The reduction in cholesterol involves the high-density lipoprotein component as well as the others.

Table 13 Changes in serum cholesterol and triglycerides during the use of the Cambridge Diet (based on the data in Di Biase et al.[44] in a trial in non-insulin-dependent diabetes)

Day	Cholesterol mmol/1)	Triglyceride (mmol/1)
0	5.51	2.40
9	4.50	1.41
26	4.36	1.27
40	4.56	1.22
54	4.64	1.30

Blood pressure

Hypertension is a fairly common feature in those who are overweight, and a fall in the blood pressure occurred after a short period on the Diet (Figure 6) even before the loss in weight was pronounced[43]. This may be due to the diuretic action of the Diet. In several of the patients who were studied, achieving a normal weight also resulted in a return to a normal blood pressure.

Diuretic effect

The Diet, used as sole source, produced a pronounced diuresis, particularly during the first few days. This is achieved without any imbalance in the electrolytes[42]. This means that patients currently receiving diuretic therapy will need at least a reduced dose, and may perhaps have the diuretic stopped completely. Indeed in the United States most of the reported adverse reactions occurred in patients who were receiving the Diet *and* diuretics. Practitioners are therefore advised to discontinue diuretics during the early stage of use of the Diet, and then to reconsider the need for their reintroduction and the dose. This is considered in more detail in Chapter 6.

Improvement in type II diabetes

For a long time it has been known that obesity may be associated with a non-insulin-dependent diabetes which may be treated by an appropriate diet with or without oral antidiabetic agents.

It is therefore not entirely surprising that the Diet as sole source has been shown[44] to correct the blood sugar (Figure 7) and reduce the high blood lipids (cholesterol and triglycerides) in type II diabetics. In such patients the oral antidiabetic agent should normally be discontinued while the Diet is being used as sole source. Once the target weight has been reached the need for reinstituting oral antidiabetic agents can be reassessed.

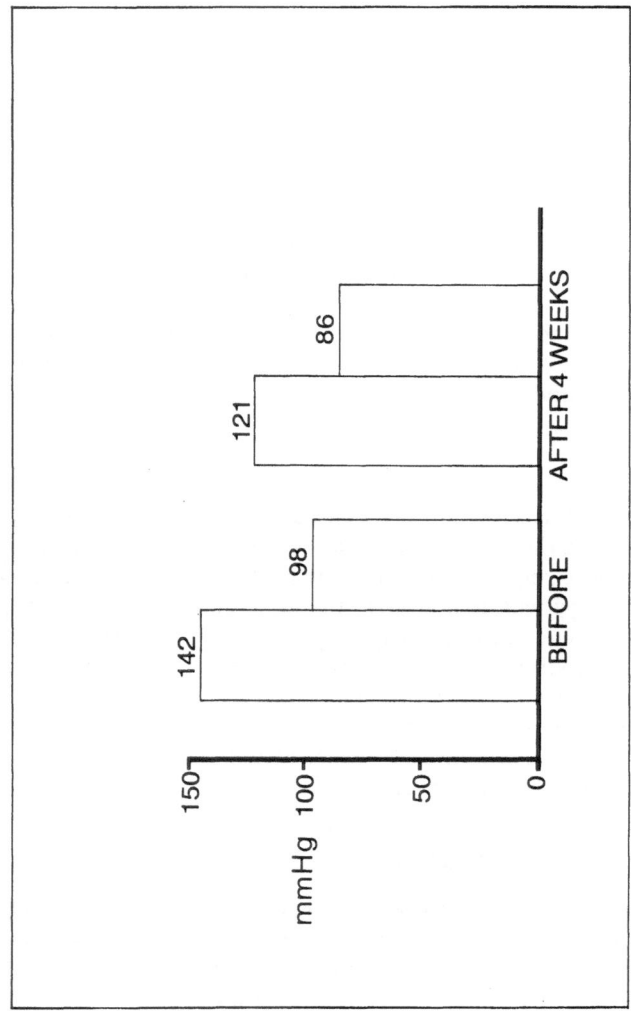

Figure 6 Changes in blood pressure during use of the Cambridge Diet. (Based on data in Kreitzman et al.[43])

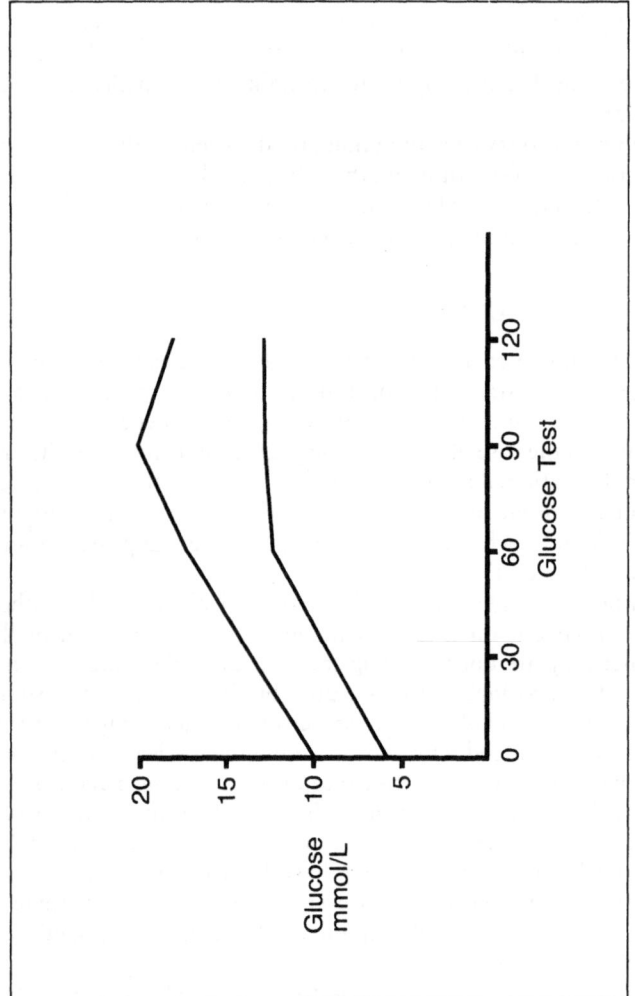

Figure 7 Influence of the Cambridge Diet on the blood sugar tolerance curve in non-insulin-dependent diabetes. (Based on data in DiBiase et al.[44])

THE NEED FOR SOMETHING TO CHEW

Some dieters experience an intense need to chew something during the use of the Cambridge Diet as their sole nutritional source, and do not find the fluid formula totally satisfying, even though they are not experiencing hunger. For such people the meal bar, which is very chewy, is very useful as a component of the daily intake. It may be taken as one meal, but many prefer to utilize it in smaller portions over the day.

It is important to try to avoid taking food – even small amounts of fruit or root vegetables. Although the calorie intake may appear to be small, the carbohydrate which they provide will reduce the ketosis, leading to a feeling of hunger and also fluid retention.

THE LAST FEW POUNDS

It is a common experience for many people to plateau just short of their target weight. Many of them start off following any diet (and the Cambridge Diet is no exception to this) correctly and experience initial good results. In the case of the Cambridge Diet these initial results are often of a dramatic reduction of weight. They begin to look and feel better and are proud of their progress. They also receive support from their friends, who notice the change in their appearance and compliment them on this fact.

Then when they are about to achieve their ideal weight they either plateau or even put on a small amount of weight. This normally implies that they are not following the Diet as they should be. This plateau is often accompanied by excuses, but it appears that in some there is a positive psychological resistance to quite achieving the target weight. In others, boredom or social commitments lead to relapses. Once the balance is broken ketosis disappears and hunger returns.

How far this is a feeling that there should still be something to aim at, and how far it is lack of motivation on the part of the dieters, who are now much more slim than they have been for many months or years, is difficult to determine. However, whatever the reason, there is no doubt that most people who slim find that the last few pounds are the most difficult to remove.

Ideally the dieter should continue the Diet to about 4 lb (2 kg) below the target weight. Once eating is resumed the glycogen stores are replaced and for each gram of glycogen 3 g of water are retained. Thus when eating is resumed most people find that they gain about 2–4 lb (1–2 kg) in weight over the first few days as a result of this fluid retention.

6

The Cambridge Diet: medical aspects of its use

In the previous chapter we noted that a substantial proportion of those undertaking any form of dieting do so without specific reference to their practitioners.

A recent survey[40] has indicated that despite the fact that the Company strongly advises that all potential dieters should consult their practitioner, 70% use the diet without so doing. Nevertheless a substantial proportion of those who are otherwise normal (apart from being overweight) and all those who require medical monitoring, are dieting under direct or indirect control of their practitioners. It is therefore important to detail specific medical aspects of the use of the diet.

THE MEDICAL CONSIDERATIONS PRIOR TO INITIATING THE USE OF THE DIET

As with any consultation concerning the question of weight loss, the important decisions to be reached are:

1. Is the person overweight (see Table 1 for the values that are usually accepted as reasonable relative to the height)?
2. If the person is overweight what should be defined as the target weight? This should take into account not only the figure defined in the table, but the person's present age and weight history. There are often merits in putting the *initial* target weight about 7–10 lb (3–4.5 kg) above the defined optimum, particularly if they have been considerably overweight for a prolonged period. Not only is this target easier to achieve, but the patient will often feel more comfortable initially at this somewhat higher level, regarding themselves as being 'too skinny' at their standard weight. If considered medically desirable the lower target can be discussed a few months later.
3. Are there any medical contraindications to the use of any form of strenuous dietary procedure (page 53)? Although these are rightly

expressed as absolute contraindications in the literature published by Cambridge Nutrition, the ultimate responsibility lies with the patient's practitioner, who will consider the overall needs of the individual patient. Nevertheless, any severe diet should be used in these conditions only after careful consideration. Other medical conditions where precautions may be required are shown in Table 15, page 55).

In addition to the questions relating to the use of the diet for patients with various medical disorders, practitioners need to consider whether any medication that is being given to the patient will either contraindicate the use of any severe diet or whether dose modification is necessary. The relevant information is given on page 60.

In consequence of these considerations, the Company instructions are that the Diet may not be supplied other than with the signed authority of the patient's practitioner in the following circumstances, to quote from the Company instructions:

"Counsellors should note that certain people should not take the Cambridge Diet as their sole source of nutrition:

1. those who have suffered recent myocardial infarction,
2. those who have recently had a stroke or major heart attack,
3. diabetics who use insulin,
4. pregnant or lactating women,
5. children below the age of 12.

People taking any of the following medication must consult their doctor, who may decide that the Cambridge Diet is unsuitable (taken sole source), or if they agree, may decide that medication needs to be stopped, or the dosage altered.

1. diuretics (water tablets),
2. antihypertensives (for high blood pressure),
3. insulin (taken by diabetics),
4. oral hypoglycaemics (taken by diabetics),
5. steroids.

For any of the people who fall into one of the categories above it is essential to have the doctor's consent before commencing the Cambridge Diet.

Explain the background and use of the Cambridge Diet to the patron in the usual way. Issue the patron with the 'Company's instruction booklet', as well as a copy of 'Information for Doctors' and a patron record card. The patron should take the information along to the doctor, and ask him or her to sign the patron medical card. Only

supply products to such people after they return the signed patron medical card."

MEDICAL CONTRAINDICATIONS TO ANY FORM OF SEVERE DIETING

As with any weight loss programme, the Cambridge Diet should be used as the sole source of nutrition only by those in good health or under the direct responsibility of the patient's practitioner.

In the conditions shown in Table 14 any form of drastic dieting (including the use of the Cambridge Diet) is contraindicated.

Table 14   Contraindications for any form of severe dieting

Type I diabetes
Recent myocardial infarction
Recent stroke or major heart attack
Pregnant and lactating women
Children below the age of 12 years

In the literature published by Cambridge Nutrition the contra-indication is rightly expressed in dogmatic and absolute terms. On the other hand the ultimate responsibility lies with the patient's prac-titioner, who will be best able to judge the overall needs of the individual patient. It is, however, advised that the Diet should be used in the conditions listed in Table 14 after very careful consideration.

In order to assist the practitioner in his decision, the possible problems involved are as follows:

1. *Type I (insulin-dependent) diabetes.* The maintenance of metabolic stability in these patients depends upon a careful balance between the intake of the individual dietary components and the insulin intake. Disturbing that balance, particularly when it involves the utilization of tissue fat, would upset that balance. Moreover insulin-dependent (type I) diabetics are rarely overweight.
2. *Recent myocardial infarction* is associated with cardiovascular instability, cardiac muscle damage and usually imbalance of body electrolytes. It is therefore essential to wait until at least several months (say 6 months) after a cardiac infarction before any form of severe dieting should be contemplated.
3. *Recent cerebrovascular accident.* This also implies an unstable cardiovascular system and the same proviso applies as to a cardiac infarction.
4. *Pregnancy and lactation.* During pregnancy and lactation, the body requirements for protein, minerals and vitamins are increased. Moreover any interference with the normal way of life is now

considered undesirable. Dieting during pregnancy is therefore contraindicated. The Cambridge Diet can, however, be used as a nutritional supplement during pregnancy. If dieting is considered essential during lactation, the four servings of Cambridge Diet may be taken, each serving reconstituted with 6 fluid ounces of skimmed milk rather than water. This will provide approximately 70 g of protein per day, just above the UK RDA for lactating women, together with the RDA's for vitamins and minerals in a calorie intake of about 800 calories.

5. *Children.* Our advice is that the Diet should not be used for weight reduction before the age of 12, but it should also be used with caution over the following 2–3 years if rapid growth is taking place. The Diet contains adequate protein when growth is not a factor, but will not allow for the additional needs for growth. If dieting is essential in early teenage the modified diet advised for lactating women should be used, but only under strict medical control. The Cambridge Diet can, however, be used as a nutritional supplement for children of all ages if one is required (see Chapter 10).

The practitioner may also wish to consider the advisability of using the diet in other medical conditions, or any precautions that should be taken in such conditions. These are shown in Table 15.

FOLLOW-UP WHEN PRACTITIONERS ARE CONTROLLING DIETING

It is suggested that the patient should be seen at the end of the first week of dieting and preferably after only 3–4 days. It is during the first few days that some patients experience difficulties, and support at the third to fourth day is helpful. The possible side-effects that may be encountered are detailed on page 65, which also describes the procedures for their reduction or relief. The main points to notice during the early stage of dieting are:

1. *Dizziness*: may be due to postural hypotension, and can be corrected by a higher fluid intake. May also be due to hypoglycaemia; in which case advise six half-portions spread over the day.
2. *Nausea and constipation*: usually due to low fluid intake, stress the need for high fluid intake.
3. *Headache*: possibly due to low fluid intake, or caffeine or carbo-hydrate withdrawal; increase fluid intake and advise simple analgesic.
4. *Hunger*: usually disappears after 2 days as mild ketosis occurs; make sure that minor 'cheating' is not the cause, and encourage continuation of diet as sole source.

Table 15 Use of the Cambridge Diet in different medical disorders (the Cambridge Diet is suitable for the following patients with the agreement and supervision of their doctors)

Disorder	Effect of the Cambridge Diet	Action
Cardiovascular disease	Weight loss will be beneficial for overweight patients, cholesterol and coagulation factors diminished	Normally suitable if patient in stable condition
Angina	Weight loss will be beneficial for overweight	If condition is stable the Cambridge Diet may be used
Gout	May precipitate an attack	Medication should be continued
Kidney disease	Not contraindicated; diet contains 1.5 g sodium, 2.0 g potassium and 34.0 g protein per day	Take dietary content into account when deciding on the medical management; should not be attempted where kidney function is known to be impaired
Heart attack more than a year ago	Not contraindicated	Diet to be taken under medical supervision
Intermittent cardiac arrhythmias	Not contraindicated	Diet to be used only under strict hospital control with possible cardiac monitoring
Rebound hypoglycaemia	Not contraindicated	Six half portions of diet to be taken per day
Depression	Not contraindicated. Weight loss could have beneficial psychological effects for overweight patients	
Lactating women	The Cambridge Diet is unsuitable as sole source of nutrition	Four servings of Cambridge Diet may be taken, each serving reconstituted with 6 fluid ounces of skimmed milk rather than water; this will provide approximately 70 g of protein per day (just above the UK RDA for lactating women)
Over 65s	Not contraindicated for those in good health	Doctor to advise
Adolescents	Not contraindicated. No evidence that it leads to anorexia nervosa	Doctor to advise; ensure patient is not anorexic before commencing; does patient really need to diet?

It must, however, be stressed that the vast majority experience no difficulties. Most will have already lost about 4–5 lb (1.8–2.2 kg) (mainly fluid) by day 3. Hence the 3–4 day visit is not essential, but is desirable.

After the 3–4 day or 1-week visit we believe that patients should be followed up at intervals of about 2–3 weeks until the target weight is achieved, though if this is difficult at any stage less frequent monitoring is acceptable if no problems have been experienced. The main purposes of these routine follow-ups are to:

1. monitor the weight loss and ensure that it is progressing normally – for average normal weight loss see page 34; for plateau see below.
2. encourage continuation of the diet either as sole nutritional source or supplemented.

For those patients whose only problem is excess weight, questioning about well-being is all that is necessary. For those with other disorders, further questions or clinical monitoring can be undertaken according to the disorder, e.g. blood pressure measurement, urine sugar in type II diabetics etc.

No laboratory monitoring is necessary unless there is suspected pathology.

For patients suffering from chronic illnesses as well as obesity, the follow-up consultations should clearly also monitor the disorder and undertake control of the medication which, as is explained on page 60, may need to be modified during the use of the diet. Side-effects which may occur later during dieting are detailed on page 65.

At the follow-up consultations when the target weight is being approached, the whole question of future diet modification and exercise should be discussed. The advised approach to the whole maintenance programme is discussed on page 80 *et seq.*

ATYPICAL RESPONSES TO THE DIET

The Plateau

Some of those who follow the diet rigidly show a steady fall in weight each week, but a substantial proportion find that weight levels off from time to time during the dieting process. This is known as 'plateauing'. While it can occur occasionally in men, it is more common in women. This is a natural process of normal life, as people who weigh themselves regularly can testify. It is mainly due to fluid retention. There are a number of possible causes, of which more than one may operate in any one person.

Perhaps the main cause is a rise in fluid retention during the premenstrual phase. Second, either through boredom with the diet or as a prime feature of the original obesity, patients may indulge in a snack. Not only does this increase the hunger feelings by reducing the level of ketosis, but it may lead to a deposition of glycogen which increases

local fluid retention (1 g glycogen holds 3 g water). The snacks may also bring the dietary intake close to the metabolic rate in those with a highly efficient metabolism.

A third cause is too high a use of carbonated diet drinks. Many of these contain sodium and hence increase sodium levels, causing possible fluid retention.

Usually the plateau only lasts for 2–4 days and a substantial weight loss follows over the following week if use of the Diet is maintained. However, if the plateau persists it may be useful to give tri-idothyronine (20 mg t.i.d.) for a week or so to stimulate further loss (page 59).

The poor responder – causes and investigation

A *very* small proportion of those who have taken the Diet have had a disappointing response. A recent survey suggests that this proportion amounts to about 1 in each 20 000 who have used the Diet. There are four main reasons for a disappointing fall in weight:

1. In the early stages the onset of the use of the Diet in women may coincide with the premenstrual stage of the cycle. Fluid retention due to the high steroid levels counteracts any loss of fat due to the Diet. This often leads to a low fluid intake leading to hyperosmolarity nausea. The dieter then gives up dieting, disappointed with the failure to achieve weight loss within the first week. Enquiry about the stage of the cycle, explanation about the cause, and encouragement to persist, will lead to rapid improvement in this group.

2. Some people find it impossible to maintain a dieting programme and yet are unwilling to admit it. In this group there is usually an initial fall of about 3.5 lb (1.5 kg) during the first week and then no further change. Careful questioning usually *fails* to elicit the extent of the food supplementation that is taking place *unless* the questioner is prepared to devote a lot of time to the session. A discussion is then necessary on whether dieting is acceptable at all. As with all diets, results are only achievable if the will exists.

   It is useful to keep a note of the quantity of Diet which is purchased, and the time period that it lasts. This is a useful indication of the dieting pattern at a later stage, particularly if the pattern of purchase changes.

3. We have seen quite a substantial number of people (probably an exaggerated number because we tend to be asked to see the poor responders) who quite inadvertently are consuming substantial calories and/or mineral in their intake of fluid. The fluids that are involved are shown in Table 12 (page 45). In our experience milk, *particularly skimmed*, natural fruit juices and alcohol are the most

common sources. Dieters often fail to appreciate that such drinks may provide calories.

4. True metabolic problems occur. We find it difficult to assess the true proportion that fall into this category in the general population because they tend to be referred specifically for investigation. All those who have the willpower to maintain the Cambridge Diet should lose weight. Arrangements have been made to study in a specialist centre the very small number who have an inexplicable poor weight loss.

Investigation of the refractory weight loss patient

The initial stage is a full history followed by clinical examination. Particular attention should be paid to the stage of life at which the patient first put on weight – did it occur during childhood, follow pregnancy or an illness, or develop gradually; are there any other symptoms which might suggest an endocrine disorder? The eating pattern needs to be discussed, and an attempt made to assess which category the patient best fits (page 18). The clinical examination should also pay particular attention to the manifestations of endocrine disorders, with special attention to the signs of clinical hypothyroidism.

In our opinion the next, and most valuable, stage is a 3–week test with the Cambridge Diet used as the sole nutritional source, with a morning and evening weight record each day and with a ketone test which they perform on the urine each morning. We make sure that the patient understands exactly how the Diet should be taken, and stress the point that during the test period they should avoid any 'nibbling' or consumption of calorie-providing drinks. We ask them to note any occasion on which they 'fall', but point out that the urine test should indicate it.

So far we have undertaken this study on 20 patients with reputed poor weight loss. In the original study on outpatients[37] the average weight loss at 1 week was 2.7 kg and at 3 weeks 5.7 kg, although the range of weight loss was large (page 34). In the present series the averages were 2.1 kg and 4.0 kg respectively. However, it appears that this group can be divided into two fairly well defined subgroups:

1. Fifteen patients with a weight loss over the 3 weeks of 3 kg or more. The average weight loss in this group is close to that of the previous study, and it appears that these probably represented a group who were not previously keeping rigidly to the diet.
2. Five patients who showed a fall in weight over the 3 weeks of less than 3 kg. This is the group that requires further study – see above.

The first stage of this subsequent laboratory investigation should be an accurate basal metabolic rate determined by the free flow oxygen (ventilated hood) method, coupled with the determination of the thermogenic response to eating.

This should be related to the mass of the subject, or more specifically to the lean body mass. We are currently using two different methods, the ^{40}K technique and a tissue electrical conductivity method to define the lean body mass. Investigations are not complete, but it appears that there may be two classes of people in the true refractory group.

1. A tiny group who are metabolically efficient and who do not show the normal metabolic response to food.
2. Another group who retain fluid on the diet. With further study we hope to be able to define these subgroups more clearly, and suggest methods by which the problems may be corrected.

Side-effects

These have not been a major problem, and are considered in detail in Chapter 7 (pages 62 *et seq*).

Consideration of the use of thyroid hormones during the use of the Diet

As the weight reduces during dieting, the metabolic rate falls (Chapter 2, page 10). This fall may start from the relatively high level of those who are overweight as a result of excessive eating (page 17) or from the lower level of the small proportion who are overweight mainly because they are metabolically efficient (page 17).

For example, after about 4 weeks on any efficient diet there is a drop in the basal metabolic rate of the order of 20%. This can lead to a fall in the rate of weight loss, or be one factor in a plateau which occurs in a small proportion of dieters. After 12–16 weeks on a diet, many patients find that the weight loss slows to about half the level which applied over the first week. It has been shown that the reduced metabolism is due to a decrease in the levels of the thyroid hormones and of their receptors in the cells[45]. Decreased circulatory levels of triodothyronine (T_3) are found.

In one controlled study[45] it was shown that a group of patients treated with T_3 20 μg three times a day after 12 weeks of dieting, when they had reached a plateau, experienced a significantly greater weight loss subsequently. This coincided with an improvement in their circulating T_3 levels.

The use of supplementary T_3 is a matter which should be decided by individual doctors if they are monitoring their patients directly. It

is suggested that the use of T_3 can lead to a loss of protein. For the vast majority of dieters it is not necessary, but when there is a pronounced and prolonged plateau it may be useful to administer it for a few weeks.

Since the thyroid releasing hormone (TRH) level of the hypothalamus and the thyroid stimulating hormone (TSH) level of the anterior pituitary are influenced in a negative feedback fashion by the T_3 level, leading to a reduction in the thyroxin level, prolonged use is not advised, and the T_3 should probably be withdrawn slowly to allow the normal body compensatory mechanisms to apply.

Effect of the Cambridge Diet on medications

The Cambridge Diet, like all very low calorie diets, and the weight loss which it produces, modifies the clinical state (page 46). In consequence of this the medication which the patient is receiving may have to be stopped, or the dosage changed.

The medication which is effected, and the suggested action, is detailed in Table 16. Specific attention is directed to the desirability of reducing the dose or stopping diuretics and antihypertensives. Clinical experience in the USA indicated that many of the side-effects which required clinical attention were the direct result of continuing the dose of one or both of these medications at the previous level.

We know of no interactions involving medications other than those detailed on Table 16. Specifically the concurrent use with tricyclic antidepressants or MAOIs is not contraindicated.

Table 16 Effect of the Cambridge Diet on medication. The Cambridge Diet, like all very low calorie diets, and the subsequent weight loss, modifies some conditions, which means that medication may have to be stopped or dosage changed

Medication	Effect of the Cambridge Diet	Action
Diuretics	Diuretic effect of the Cambridge Diet would give rise to excessive potassium loss	Diuretic medication can be stopped altogether or reduced, in which case potassium supplement should be given with appropriate serum monitoring
Antihypertensive	Potentiates treatment	Medication should be reduced or stopped altogether when the Cambridge Diet is begun
Insulin (type I diabetes mellitus)	In type I insulin-dependent diabetes the Cambridge Diet is nearly always contraindicated because of the problems of regulating blood sugar levels, and few insulin-dependent diabetics are overweight	Weight reduction with the Cambridge Diet should only be carried out under careful and strict hospital control, and should normally be started in hospital
Oral hypoglycaemics (antidiabetic tablets) (type II diabetes)	Potentiates treatment	Medication must be reduced or stopped altogether when the Cambridge Diet is begun
Steroids	Sodium retention	The Cambridge Diet is normally contraindicated

7

The Cambridge Diet – fact and fallacy

Perhaps because the Cambridge Diet is revolutionary, it has been the subject of controversy on both sides of the Atlantic. It was extensively attacked by many US writers who have relied upon ex-cathedra pronouncements rather than experimental evidence. These range from the FDA[46], the AMA[47], the American Dietetic Association[48], the American Society of Bariatric Physicians[49] to eminent university nutritionists and doctors[50], university extension services[51], science writers, health authorities[52] and journalists[53].

The medical literature abounds with examples of vicious attacks on new developments that conflict with current medical dogma, and the Cambridge Diet is no exception to the rule. As has so often happened in the past, the most vociferous antagonists have sometimes converted themselves into protagonists by the simple process of personal experience. It is therefore important to consider the medical facts and fallacies that are related to the Cambridge Diet.

THE RELATIONSHIP OF THE CAMBRIDGE DIET TO THE 'LIQUID PROTEIN DIET'

In the United States in 1977–78 a medical scandal took place, the repercussions of which are still influencing the developments of very low calorie diets.

In the 1970s George Blackburn[27,28], developed his 'protein-sparing modified diet' which consisted of a daily intake of about 100 g lean steak with vitamin and mineral tablets. Almost simultaneously a method was discovered of extracting the proteins from cowhide. This extract was sweetened with saccharin, flavoured and produced as a cheaper and more convenient form of the protein-sparing diet under the term 'liquid protein diet'. This 'liquid protein diet' was subjected to no clinical trials before it was sold in the USA.

The 'liquid protein diet' contained very poor-quality protein – equivalent to gelatin – deficient in several of the important amino acids. It contained no other nutrients and physicians were required to prescribe

the necessary minerals and vitamins separately. Not all patients were given, or took, these supplements. The total absence of carbohydrate gave rise to a severe ketosis and loss of potassium.

At the end of 1978 several deaths were reported in patients who had been taking the 'liquid protein diet'. Eventually over 50 deaths were reported, of which approximately a third could be directly attributed to the use of the diet[54,55]. Autopsies revealed severe degeneration of the heart muscle, which correlated with abnormal heart rhythms and electrocardiographic abnormalities found during the terminal stages. These abnormal rhythms could not be corrected by therapy and are of the type seen with total starvation[56].

There have been several theories about the causation of these heart abnormalities. These have included a general loss of protein, absence of specific amino acids, imbalance of electrolytes (particularly potassium deficiency) and toxicity of the preparation. To date there is no clear knowledge of the exact cause of the deaths, all of which occurred several months after the patients started the 'liquid protein diet'.

It is important to appreciate that there are several vitally important differences between the Cambridge Diet and the 'liquid protein diet', despite the fact that medical and lay writers have erroneously linked the two.

First, whereas the 'liquid protein diet' contained only one inferior protein nutrient lacking in several of the essential amino acids, the Cambridge Diet contains a protein source with all the essential amino acids in adequate quantities (Table 9, page 41): carbohydrate; essential fatty acids; 13 vitamins; 14 mineral trace elements (Table 10, page 42) and fibre. For all those substances for which there is a USA-recommended daily intake, the Cambridge Diet contains at least this quantity. The USA RDAs are more extensive than those in the UK. Hence UK requirements are adequately met.

Second, the Cambridge Diet was subjected to extensive clinical study in many research units in several countries over many years (Table 8, page 36), whereas the 'liquid protein diet' was marketed without any rigorous premarketing testing.

Third, the morbidity and mortality became apparent by the time the 'liquid protein diet' had been used by well under 1 million people. In contrast the Cambridge Diet has already been used by some 7 million people, of whom approaching 1 million have used the Cambridge Diet as the sole source of nutrition for at least 4 weeks. This extensive usage, monitored for side-effects and toxicity both in the United States and the United Kingdom, has confirmed the safety of the Cambridge Diet which had been established during the clinical trials.

Fourth, the electrocardiographic manifestations of starvation and low potassium are now well known[56]. Extensive electrocardiographic

monitoring has been undertaken during the clinical trials of the Cambridge Diet, and these changes have not been encountered[39,42,43].

'CAMBRIDGE DIET: MORE MAYHEM'[57]

This is the type of headline which appeared over an article critical about the Cambridge Diet in the *Journal of the American Medical Association*[57], and it is therefore important to examine the evidence[58].

The extent of the clinical studies that have been published on the Cambridge Diet has already been described in Chapter 4, and is shown in Appendix 1. These studies have not only examined the safety of the Diet in terms of routine clinical observations over short- and long-term use, but have also relied upon electrocardiographic evidence. In these clinical trials the safety of the Diet was clearly established. However, recent medical experience of pharmaceuticals has demonstrated that even after extensive and competent clinical trials an unacceptable level of toxicity may occur during routine postmarketing use which can give rise to withdrawal of the product.

The most extensive postmarketing investigation of the Cambridge Diet was that undertaken by the Food and Drug Administration (FDA) in the United States[46].

Between the first commercial appearance of the Cambridge Diet in the United States in March 1980 and November 1982 it was calculated that some 3 million people had used the diet. Of these about 850 000 used it as their sole source of food for 4 weeks or more. The FDA *invited* reports of side-effects or deaths, and six deaths were reported to them of people who had taken the Cambridge Diet. The FDA thoroughly investigated all the individual reports and found no evidence linking the death to the use of the Cambridge Diet. There was a reasonable alternative cause of death in each case. Five of them were already suffering from diabetes or coronary atherosclerosis, or came from families with a history of early coronary deaths. None of the deaths bore any resemblance to those reported with the 'liquid protein diet' with abnormal cardiac rhythm and heart muscle degeneration.

On the basis of statistical probability in a population cohort of that age and sex, natural causes would be expected to result in approximately two deaths per year per 100 000 people even if one excludes the excess mortality of obesity. Hence the death rate of those using the Cambridge Diet is less than one-third the anticipated figure.

As a result of the deaths from the 'liquid protein diet' the FDA brought in new regulations in April 1984 making it mandatory for all preparations with more than 50% of its total calories derived from protein or liquid protein, and containing less than 400 calories per daily intake, to carry a warning 'These preparations may cause serious

illness or death'. After discussion with the Company, which reviewed the safety record, the FDA did not include the Cambridge Diet in these regulations and the Diet does not carry any such warning in the USA or elsewhere.

In addition to the six deaths the FDA in their 1983 report on their investigation of the Diet referred to four hospitalizations on the Diet. All these patients had been taking diuretics simultaneously with the Diet and an electrolyte imbalance had occurred. The literature clearly warns against this (see page 52). The only other reports of side-effects among the 3 million who had taken the diet at that stage were 138 reports of minor symptoms such as nausea, diarrhoea, vomiting and headaches.

In the United Kingdom, Cambridge Nutrition is monitoring all adverse effects reported to it by Counsellors and by the medical profession. Among an estimated use which is now approaching 1 million people, no serious side-effects have yet been reported. The few minor side-effects are those already established during the trial stage, for which measures of avoidance are already known. These are detailed below.

Hence on the basis of the available information there is no evidence of danger from the use of the Cambridge Diet.

ADVERSE REACTIONS

Adverse reactions are encountered only rarely when the Cambridge diet is taken as directed, but a few relatively minor side-effects may occur during the first few days use of the diet if the directions are not followed. A list of these side-effects, together with their probable mechanism and mechanism of relief, is given in Table 17.

Dehydration

The Cambridge Diet contains a rather high concentration of minerals when made up into the volume of hot or cold water which produces a palatable drink. It is this concentrated mineral level which probably accounts for the nausea which some patients experience initially, and for some cases of diarrhoea.

The condition can be very readily avoided by drinking a full glass of water immediately after taking the Diet. The Diet produces a natural diuresis, particularly in the early stages, and this leads to a dehydration which probably accentuates the nausea and may be one factor in the genesis of the headache experienced by some subjects during the first few days. The balance of the electrolytes in the Diet is such that an imbalance from the diuresis is virtually impossible unless

Table 17 Possible side-effects and appropriate action

Side-effect	Mechanism	Action
Dehydration	Diuresis	High fluid intake
Diarrhoea	Concentrated minerals	High fluid intake and use as six small portions
	Lactose intolerance	Lactase pretreatment
Constipation	Low bulk	High fluid intake and fibre
Headaches	Dehydration	High fluid intake
	Carbohydrate or caffeine withdrawal	Simple analgesics
Nausea	Concentrated minerals	High fluid intake after diet
Mild dizziness	Dehydration	High fluid intake
	Hypoglycaemia	Divide into six small portions
Electrolyte imbalance	Diuresis + diuretics	Stop diuretics – monitor and give potassium
Gout	Uric acid rise	Give uricosuric therapy

a diuretic is being consumed at the same time (page 52). The standard advice is to withhold diuretics during the first few days of using the Diet, for they are rarely needed at that period. If, however, they are continued the practitioner should monitor the blood electrolyte levels. The likely correction that will be necessary is an increase in potassium intake.

However, the most important aspect of taking the diet is a high bland fluid intake (minimum 3 pints; preferably over 4 pints).

Headache

Apart from fluid imbalance there would appear to be two other causes of headache during the first 2 days of the diet. The first of these results from the withdrawal of carbohydrate, and is a somewhat exaggerated ketotic response. The balance of the components of the Diet is such that this is rapidly eliminated. The second is the direct result of caffeine withdrawal. The headache responds to the administration of mild standard remedies (e.g. paracetamol, aspirin).

Mild dizziness

This also is experienced by a very small percentage of people during the initial stage of the use of the Diet. It is often due to postural hypotension and is seen more commonly when diuretics and/or anti-hypertensives have been continued. It may also be due to mild hypoglycaemia. This can be corrected by taking the daily intake split into a large number of smaller quantities, e.g six half-scoops at 3-hourly intervals during waking hours.

Diarrhoea

The diarrhoea which results from the high mineral content of the Diet has already been detailed: it normally lasts for no more than 48 hours. Some other people experience continued diarrhoea on the Diet due to lactose intolerance. Since the predominant sugar is lactase from the skimmed milk powder which forms a significant portion of the Diet, a congenital absence of the enzyme lactose leads to unaltered lactose entering the colon. Here the bacteria metabolize the lactase, creating both flatulence and loose motions.

This can be corrected by the use of the enzyme lactase available as Lactaid from G. Morgan Ltd of 75 Woodbridge Road, Guildford, Surrey (Guildford 67326/7). Three doses of Lactaid are added to 1 day's made-up supply of Cambridge Diet, stirred well and stored *in the refrigerator* for 24 hours. The storage of the diet in liquid form may theoretically reduce the vitamin C content. If this procedure is used while the person is using Cambridge Diet as the sole nutritional intake, it is desirable but not essential to advise an extra 50 mg ascorbic acid intake daily.

Although this advice is given on the basis of general principles, neither of the authors has seen a case in which there is any evidence of vitamin C deficiency. The diet provides 60 mg per day which, though it is equal to the US RDA, provides a good margin above scorbutic level.

Lactaid tablets may also be taken at the same time as the Diet, though it is not clear whether this is as effective a method for the management of lactose intolerance.

Constipation

Since the Cambridge Diet contains very little fibre (this whole question of the fibre content is considered elsewhere, page 83), bowel movements are less frequent and smaller. If an adequate amount of water is drunk, the presence of the minerals in the diet normally ensures that constipation with hard faeces is not a problem. However if necessary a natural laxative (e.g. Fybogel), together with an increase in the fluid intake, will solve the problem. It is preferable to avoid dietary fibre preparations (e.g. proprietary brans) which often have a high calorie content. On very rare occasions it has been necessary to give Epsom salts intermittently.

Gout

Precipitation of an attack of gout is a theoretical possibility and has been reported rarely. When used as sole food source the Cambridge

Diet causes a slight increase in uric acid during the first 2 weeks. For the majority of people this increase is of no consequence and the uric acid levels falls again. However, in a patient who suffers from gout a brief attack may be precipitated and any uricosuric agents (e.g. allopurinal) should be continued when the diet is given.

Halitosis

This is a direct result of the mild ketosis, which is a vital component of the effective use of the Diet. It is worse if the fluid intake is too low. The situation can be improved by the frequent use of mouthwashes or anti-halitosis sprays.

Fatigue

About 2% of patients complain of tiredness at some stage of the Diet. The vast majority subsequently recover spontaneously – probably as the mild ketosis develops. We have occasionally advised a fourth helping, particularly if both tiredness and hunger are present.

Hunger

This is a fairly constant feature of just the first 2–4 days of the diet until the mild ketosis occurs. Day 3 is usually the worst day and patients should be advised about this in advance (page 54). A high fluid intake and a busy work schedule, which does not allow time for concentration on food, helps. Very rarely hunger persists for a few more days. If so it is better for the dieter to take an extra Diet meal, than to nibble, so that the balance of the ingredients to produce the mild ketosis is less disturbed (see also page 70).

Flatulence

Flatulence may be an uncomfortable feature, particularly associated with the use of the nutritional bar. It is due to the partial breakdown of one of the constituents by the bacteria in the colon. Though it is an uncomfortable feature, it has no medical relevance.

Thus it is apparent that there are several minor side-effects which may occur on the Diet, particularly during the first few days. The majority of these can be avoided by an adequate fluid intake of *at least* 3–4 pints per day *in addition* to the 1½ pints used for the preparation of the Diet meals.

Even if they occur, the side-effects disappear very rapidly and the success of the Diet in terms of weight reduction is such that the side-

effects are readily tolerated.

A few side-effects have also been encountered later in the dieting stage:

1. *Feeling cold.* As the weight falls the lean body mass is reduced by about 30% of the weight which is lost[17] (page 70). Hence the resting metabolic rate falls. Coinciding with this the amount of subcutaneous fat is reduced and heat loss is increased; hence the patient may complain more about feeling the cold.

   Neither of the authors has heard of the use of the Cambridge Diet giving rise to clinical hypothyroidism. However, there is a possibility that if a mild hypothyroidism was one factor in the genesis of the obesity, the reduction in metabolic rate which results from the lower lean body mass could accentuate the clinical condition. Practitioners should therefore consider the possibility of hypothyroidism if excess sensitivity to cold becomes a symptom.

2. *Hair loss.* Loss of hair can occur without known cause, and is a feature of both hypothyroidism and weight loss (perhaps in consequence of the fall in metabolic rate). To date we have heard of six cases of some hair loss (none of them serious). All occurred after prolonged successful dieting.

   In each we advised the addition of three capsules of a vegetable oil per day (to provide extra essential fatty acids above the recommended level in case this was a factor). In no case are we convinced that this was necessary, but like all such hair loss, regrowth occurs on a maintenance diet.

3. *Menstrual irregularity.* Any weight change can be associated with menstrual changes – mainly irregularity. Such changes can occur during the use of the Cambridge Diet but have no clinical significance.

4. *Allergic reaction.* As with any food, a small minority of people may show an allergic reaction. This may be an allergy to milk protein, in which case the Cambridge Diet cannot be used. Usually, however, it is to one ingredient of a particular formulation and a change to a different flavour clears the problem.

5. *Fluid retention.* This is considered under 'plateau' on page 56.

KETOSIS ASSOCIATED WITH THE DIET

During starvation a severe ketosis is inevitable after the first day or so as the body uses the stored fat as the metabolic fuel without an adequate supply of carbohydrate or glucogenic amino acids to enable full metabolism of the ketone bodies via the Krebs (citric acid) cycle.

As the amount of carbohydrate is increased in any diet the level of ketosis is reduced. The amounts of carbohydrate in the Cambridge

Diet are such that a mild ketosis is present after the first day or so when the Diet is used as sole nutritional source.

Whereas a severe ketosis is both unpleasant and medically undesirable because the beta-hydroxy butyric acid and aceto acetic acid result in an acidemia, a mild ketosis has no such undesirable effects. Indeed the majority experience only minimal hunger once the ketosis has occurred (after about 48–72 hours) and feel better due to the euphoric effect of the mild ketosis.

Those who 'cheat', and supplement the Diet with small snacks, do not develop the ketosis and hence, by virtue of increasing their intake, experience even greater hunger, a self-perpetuating problem.

For those who experience the need for dietary intake, the best advice is to avoid snacks but to take a fourth helping of the Diet, which avoids the problem.

THE NITROGEN BALANCE AND LEAN BODY MASS

One of the main objections made against any very low calorie diet with its protein content lower than the normal intake is that it gives rise to a negative nitrogen balance and hence to a fall in the body protein – one of the main components of the 'lean body mass'. By implication this protein loss will come at the best from the body muscle and at the worst from the heart muscle, linking the situation to the cardiac muscle degeneration caused by the 'liquid protein diet' (page 62).

During normal dieting there is no way that lean body mass will not be lost *whatever the method* – if weight is lost lean body mass will automatically be lost.

Lean body mass is defined as 'total weight of the body minus its fatty tissue'. Essentially it consists of the skeleton, the organs and the protein component of the body including the muscles. *It also comprises the non-fatty component of the fat-containing tissues.*

Using a radioactive potassium (^{40}K) technique James et al.[17] showed that overweight people have a higher lean body mass than do lean people. Approximately 30% of the additional weight of an overweight person is lean body mass and of this about 25% (i.e. 9% of the total) is protein – probably predominantly the structural components of the tissue cells in which the fat is deposited.

By whatever method this excess weight is lost the lean body mass component of the fat also goes. This releases a substantial buffer of protein and can fully account for the additional nitrogen in the balance equation during the early stages of the Cambridge Diet (Figure 8).

In a study it has been possible to demonstrate that, as the weight falls on the Cambridge Diet, the proportion of the body as lean body

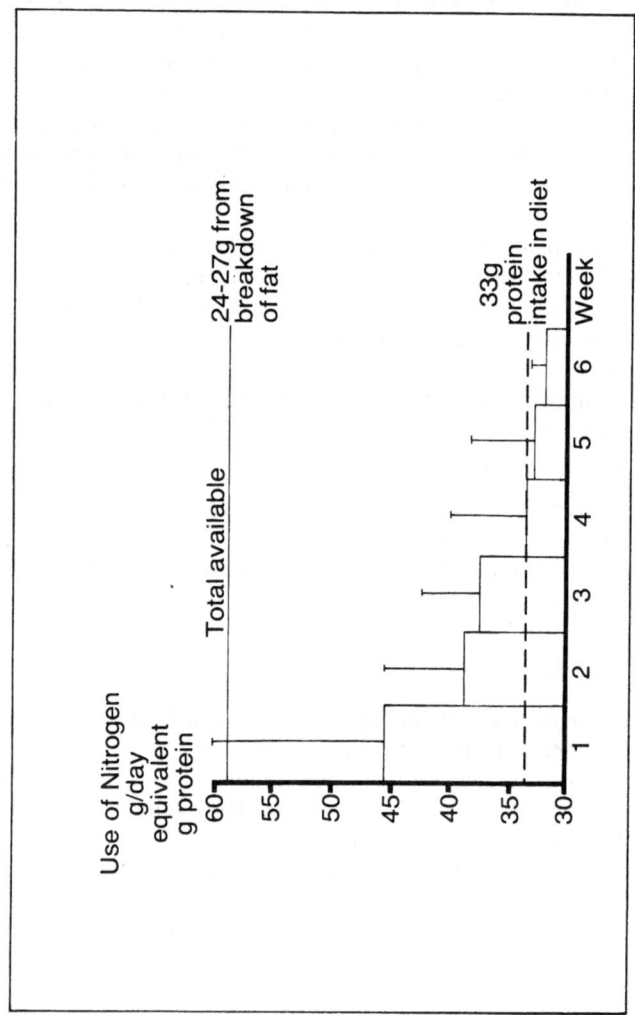

Figure 8 The balance between nitrogen needs of the body and their supply from the Cambridge Diet and from release during loss of 'fat tissue'

mass measured by the ^{40}K method remains unchanged[59]. This is exactly what would be anticipated from the observations of James et al.[17]. There is no evidence that any protein is lost from the body muscles or from the heart muscle.

In a recent study in Cambridge it was found that there was no loss of performance during a rigorous training programme among the University Women's Lightweights Squad, while they were reducing to the defined weight using the Diet. Indeed, in a study currently in preparation the effect of weight training exercises is to be examined during fat reduction on the Diet to determine whether it is possible for the muscle mass to be increased utilizing amino acids freed from the adipose tissue.

Equally, as explained previously, there is no evidence that the heart muscle is depleted (page 63). No cardiac irregularities have been determined, and electrocardiography – both random and continuous – has shown no abnormalities of the type seen with either the 'liquid protein diet' or starvation[56].

It is also important to appreciate that the actual determination of the nitrogen balance suffers from inherent errors[60], which may tend to suggest a negative balance when none exists:

1. inherent errors in the analytical methodology,
2. all the food intake may not be consumed,
3. there may be errors in collection if the total excreta and changes in skin, hair, mucus, etc. are not taken into account.

THE DIET AND ANOREXIA

It has been suggested that dieting in general can lead to anorexia nervosa. Whether it can do so is medically disputed.

Equally it has been suggested that the anorexia associated with the mild ketosis during the use of the Cambridge Diet would be an added factor in the genesis of anorexia nervosa.

We have undertaken a careful monitoring for the possibility of anorexia nervosa caused by the use of the Cambridge Diet. There is no such evidence during the extensive trials and clinical experience in the USA and the UK. It is impossible to prove a negative correlation of this type, but we are of the firm opinion, on the basis of the clinical evidence, that the Cambridge Diet does not lead to anorexia nervosa.

8

The Cambridge Counsellor system

The distribution of the Cambridge Diet is organized via a multilevel independent marketing structure called the Cambridge Counsellors. The Cambridge Counsellor is an independent agent who purchases the Diet directly from the Company at a basic wholesale price and provides it to the dieters – the 'Patrons'. The Counsellor makes a reasonable profit on each transaction, but for this margin provides a service to the Patron.

The Counsellor's duty is to advise, motivate and liaise with the Patrons and to provide the necessary support and encouragement while they are using the Diet.

Counsellors are selected initially from those who have used the Diet successfully themselves. Hence they have direct personal experience of the use of the Diet and are in the ideal position to pass this experience on to others. Those who have successfully used the Cambridge Diet want to share their enthusiasm and success with others. It is virtually impossible to sell any diet if you are overweight, or at least have not lost a substantial amount of weight.

Their selection is initially based on the advice of an existing Cambridge Counsellor but before they are accepted by the Company each must pass an extensive questionnaire examination about the product. This tests for a thorough understanding of the diet – its correct usage, how to identify those people who should not use it, and the rudiments of healthy eating. In short, while they are in no way intended to take the place of qualified medical staff in the treatment of the obese, they should nevertheless be fully equipped to counsel the otherwise healthy obese.

The educational background and previous training of the Counsellors varies tremendously, and this variation will probably increase rather than diminish with time. Current Counsellors include doctors, nurses, dieticians and doctor's wives and receptionists, as well as teachers, physical education experts and executives. Education in nutrition is not a prerequisite for selection as a Counsellor, although this would clearly be advantageous. The Company regards one of its

main duties as providing adequate nutritional education.

Initial contact with potential Patrons is still largely based upon personal recommendation, though the Company places advertisements, and Counsellors may announce their availability, using Company-approved wording, in local newspapers. Some Counsellors prefer to see potential Patrons singly from the beginning, while others find that the initial discussion of the use of the Diet can best be undertaken in a group session. At the group meeting, held either at home or in a special meeting room, the Counsellor can educate, instruct and demonstrate the virutes of the Diet and answer any questions. Many Counsellors find that it is useful to have one or more Patrons present who have successfully lost weight and can speak about their experience. Direct evidence is far more convincing than hearsay.

However, before any Patron can be provided with the Diet, Counsellors are required to complete a Patron record card (Table 18). One of the important features of this is to determine whether the potential Patron is suffering from any medical disorders or utilizing any form of medication which would contraindicate the use of the Cambridge Diet or any other weight-reducing diet (page 53).

Table 18 Patron medical card

1a. I understand that it is important for me to see my own doctor about dieting*.
1b. You are strongly advised to do so before dieting.
2. Are you taking any drugs?
    If 'YES' please give details. . . . .
3. Do you suffer with, or have a family history of:
    Kidney complaints. . . . . Asthma. . . . . .Heart disease. . . . .
    Diabetes. . . . . .High blood pressure. . . . . .Other. . . . .
4. Have you recently been hospitalized YES/NO
    If 'YES' please give details. . . . .
5. Are you pregnant . . . . Had a baby in the last 3 months. . . . .
6. Are you allergic to milk. . . . .
7. Please give any additional medical information you feel we should know . . . . . .
    . . . .

I have completed these questions to the best of my knowledge and will inform you of any changes in the above.

Signed. . . . .

---

* See Table 19.

If there are *no* such obvious contraindications the Counsellor must still advise the prospective Patrons to consult their own doctors about the use of the Diet. The record card contains a space for a signature that such advice has been received.

The Company considered very carefully whether it should insist upon a practitioner's signature before the Diet could be supplied under any circumstances, and decided that this was an inappropriate

requirement within a free society. Anybody can eat or drink whatever they decide, and can attempt to put on or take off weight by a large number of different methods. Many of these are potentially far more dangerous than the Cambridge Diet, for which there is no known danger in people who are otherwise healthy apart from being overweight.

If, however, potential Patrons are suffering from any of the disorders defined on the record card, or receiving any medications, Counsellors are instructed that they must *not* supply the Diet until there is a signed authority from the Patron's practitioner, either indicating that there is no contraindication to the use of the Diet, or that the practitioner will supervise the use directly (Table 19).

Table 19 Reverse of Patron medical card – medical approval

Note to doctors:-

Cambridge Counsellors are specifically instructed not to supply the diet to Patrons who are receiving medication from their doctor until the doctor signifies His/Her agreement.

Are you willing for:- ....
to receive the diet under your supervision?

YES/NO

Doctors signature:-................................... Date ....

When the Patron takes a decision to use the Cambridge Diet the next duty of the Counsellor is to advise on the method of use, including how to reconstitute it, the need for extra fluid, and avoiding other food and drink items containing calories or sodium.

It is our opinion that the best results are achieved when the Cambridge Diet is used as sole source of nutrition, and we believe that this is the advice usually given.

On the other hand, many people can lose weight successfully using the modified diet of three Cambridge Diet meals and one 400-calorie meal per day (730 cal/day – page 43). We know that some Counsellors advise this alternative initially. Since the full nutritional requirement (other than calories) is provided by this procedure, we have seen no reason to advise against it.

Losing weight is an extremely personal and serious project for the overweight sufferer. The dieter needs personal contact, reassurance, praise and encouragement, compliments to raise the self-image, and above all constant motivation. The Counsellor should make regular contact with the Patrons, particularly during the first 10–14 days, but perhaps even more important than this they should be readily available. Day 3 is usually the most difficult day, and contact at this time is

especially important. The Counsellor is always available to listen and to provide an honest interest in the Patron's progress.

A vital feature of most techniques that have achieved weight control has been adequate and sustained motivation. In the case of the Cambridge Diet this advice, motivation and assistance is provided by the Cambridge Counsellors.

The further function of the Counsellor, and in the long term the most important, comes when the target weight is achieved. The duty of the Counsellor does not cease at this stage. The aim must be maintenance of the target weight by sensible nutritional habits.

The Company plays an active part here – extensive educational material is supplied to the qualifying Counsellor and this is followed up by a monthly magazine which includes articles about nutrition and healthy cooking in an attractive and readable form. Training sessions run by the Company and by Group Counsellors all contribute to this education, and the end result is that Counsellors are not only well informed, but trained to pass this information effectively to their patrons.

THE TRAINING OF CAMBRIDGE COUNSELLORS

As has been explained above, previous biological or medical training is helpful but not essential to the Counsellor, for the following reasons:

1. Cambridge Counsellors should not be expected to act as physicians, nurses or general nutritionists when they are acting as Counsellors. The medical function should be performed by the Patron's own practitioner and the Counsellor should ensure that the practitioner is consulted as necessary.

   Nor should the Counsellor act as a general nutritionist. General nutritionists and dieticians have a vital role to play in society. This covers both the provision and education of those with special dietary needs (.e.g. diabetes, phenylketonuria, etc.) and also the more general education which will help to ensure that nutritional deficiencies are problems of the past, both in industrialized and Third World countries.

   There should be no conflict, nor need for conflict, between Cambridge Counsellors and dieticians. Counsellors help dieticians by assisting over one of their most time-consuming activities, releasing them for their more important and vital roles.

2. It is the responsibility of the Company to ensure that the Counsellors are trained appropriately and adequately for their full role, and this is a responsibility which the Company fully accepts. The

Company undertakes this training by various techniques including:

(a) A hierarchical Counsellor structure with more senior and experienced Counsellors instructing, advising and controlling those who have been appointed more recently.
(b) Special training officers.
(c) Written material in various forms.
(d) Conferences and small teaching sessions provided by experts in the field.
(e) Access to expert advice in the Company's office.
(f) Visual (e.g. flip charts and slides), audio and video taped material.

In order to ensure that training is progressing, the Company is establishing an examination structure through which those Counsellors who are progressing to more senior positions must proceed.

From this description it should not be thought that a fully and finitely structured training system has already been fully established, though it has been defined and substantially implemented. First the Company is young, and to develop a training programme without errors takes time. Second, the process of training itself must be a developing one which takes account of the needs as they are identified.

THE COMPANY CONTROL OF THE OPERATION OF THE COUNSELLOR SYSTEM

Experience shows that the majority of those who become Counsellors are those who have a caring nature and want to encourage others to achieve the breakthrough in weight control that they have experienced themselves. But whenever there is a financial incentive to performance it has to be accepted that, human nature being what it is, there is a possibility of abuse of the system. The Company has recognized this danger, and has established a review board of senior Company members, which meets regularly to consider reports of infringements of Company policy by Counsellors.

A series of penalties is defined, ranging from a warning letter to termination of Counsellor agreement. Complaints which may be notified by members of the public or other Counsellors include such matters as retailing the Diet in a shop, inappropriate advertising or poor quality of Counselling.

The number of complaints has, as might be expected, been small, The review board anticipate that, given the type of person who seeks to become a Counsellor and the peer selection process which is adopted, the publication of a Code of Practice, backed by publication of review board action, will maintain a good standard of Counselling.

RELATIONSHIP OF THE COUNSELLOR SYSTEM TO THE MEDICAL PROFESSION

It is our firm hope that no conflict will exist between any component of the Cambridge system and any part of the medical and allied professions. The current arrangements are designed to avoid any such conflict, for we see the Counsellor system as an adjunct to one very small but important feature of medical practice, namely the management of the overweight.

When the practitioner is interested in the clinical management of this problem then the Counsellors are more than happy to ensure that the practitioner takes the controlling role in the clinical management, relegating themselves solely to the function of supplying the diet. Indeed this is precisely the role that the Counsellors must take when either medical disorders or medications make it imperative that medical control should be a feature throughout the weight reduction period (page 53).

On the other hand, we know from many years of experience that practitioners are not generally enthusiastic about the practical management of those who, for medical or cosmetic reasons, or both, should lose weight. Many practitioners refer their obesity problems to some form of food restriction therapy programme group.

In our opinion Cambridge Counsellors provide a better service of advice, encouragement and support than some of these groups. Furthermore it is abundantly clear that the Cambridge Diet is nutritionally superior and far easier to manage than diets which require weighing and measuring by the dieter.

It is to be hoped that all prospective Patrons will accept the advice given by the Counsellors that they should see their own practitioners. This will enable the practitioner to advise on the need for dieting, suggest a target weight and also define the intervals at which he or she would like to see the patient.

When dieting other than under direct medical control, Patrons are instructed that they should take the Cambridge Diet as sole source for 4 weeks only, continuously. They should then take an extra 400 calorie meal daily for 7 days, in addition to the Diet, before returning to sole source. If, however, the practitioner is monitoring the patient's progress at intervals of, say 2 weeks, the week of higher intake need only be included at the practitioner's discretion. We have not included the 730 calorie/day week in the regime at the Cambridge clinic.

If practitioners wish to see how Cambridge Diet meetings are organized, or the information that is presented, they are very welcome to attend, with or without previous announcement. We are aware of several practitioners who have been interested, have attended and have subsequently advised the use of the diet.

SUPPLIES OF CAMBRIDGE DIET FOR USE BY PRACTITIONERS

The question is raised periodically whether medical practitioners can be supplied with diet direct by the Company for use in their practices.

As a matter of policy, to achieve a single distribution arrangement and to avoid confusion, it has been decided that there should be a single system of distribution through Counsellors. This does not, however, mean that the Cambridge Diet cannot be available for practice use.

1. The practitioner may become a Counsellor, and hence have direct access to the diet.
2. Several practitioners have arranged that their spouse becomes a Counsellor and arranges supplies through the spouse.
3. In other practices an arrangement has been made for the practice nurse to become a Counsellor, and to arrange a regular obesity clinic within the practice supervised by the practitioner.

9

Weight maintenance and healthy living

MOVING INTO THE MAINTENANCE STAGE

The change to the maintenance programme should not be a sudden process, but should be gradually introduced as the target weight is reached. Unless this stage is successfully accomplished then the dieting process will have been a waste of time and the typical 'yo-yo weight syndrome' will occur. The important consideration is to establish not only a more sensible eating pattern utilizing current nutritional concepts, but to achieve a new set point of weight by the long-term maintenance of the target weight by calorie restriction with an adequate intake of the essential components of the diet.

At the changeover stage the continued use of the Cambridge Diet is of paramount importance. Once off the diet as the sole food source, the dieter should continue to take the Cambridge Diet three times a day and supplement it with ordinary food of modest calorie content. The use of the Cambridge Diet will ensure that however low the normal food supplement, the essential dietary needs of the patients will be met. Initially it is advised that one balanced meal of 400 calories each day should be added. The doctor should stress the importance of not increasing the intake of normal food too rapidly. Firstly, if the diet has been used for any prolonged period the stomach itself will have contracted, and nausea may result from a large intake. Secondly, and more importantly, the metabolic rate will have fallen (as it does with any effective diet) and in consequence, until stability is re-established, a dietary intake lower than that which previously maintained a stable weight is essential (page 17).

The intake of a sensible, balanced ordinary diet can then be gradually increased while the target weight is maintained. One or more portions of the Cambridge Diet are taken during this process until it is clear that an adequate mixed diet is being consumed, and that there is no risk of dietary inadequacy. If this process is done slowly, and the

weight is maintained at the target for a prolonged period, experience with the Cambridge Diet indicates that a new set point can be established at the target weight. The secret is patience and persistence.

If social or business commitments interfere too much with this regime, and too many calories are perforce consumed, then the patient simply goes back to the Diet as sole source for a few days and the process of re-education can start again.

It is important that the patient should keep a regular check on weight, using the same set of scales and weighing at a regular time on each occasion. There are fluctuations in the weight of most normal people from day to day, probably depending upon water retention, and this effect is increased in the premenstrual period. Different authorities differ on whether they believe the weighing should be undertaken daily or less frequently. Certainly the interval between weighings should not exceed 1 week. There are also differing views about the change in weight that should be regarded as a reason for modification in food intake. The limit to be defined should, as far as possible, have some relationship to the degree of daily variation in the weight of the person concerned. In the early stage of a weight maintenance programme it is probably better to err on the side of defining rather too tight limits than too relaxed ones. At this stage it is probably wise to suggest that there should be a change in the food intake if the weight varies by more than 5 lb (2.2 kg) and certainly if it varies by 7 lb (3 kg) from the target level. It is much easier, both physically and psychologically, to adjust a small weight change than a large one.

Whether the patient should be advised to return to Cambridge Diet as the sole food source for a period until the weight returns to the target, or whether the change should be dealt with by a modification of the amount or content of the normal food intake, will clearly depend upon the stage of the maintenance programme which the patient has reached and the rate of the weight rise. So far as possible it is desirable to concentrate on the aspect of rational sensible eating to adjust minor changes since a return to full sensible nutrition should be the aim in all patients.

Essentially, change to a weight maintenance programme must be a matter of trial and error within a framework of the principles of sensible nutrition. The variation of needs is too great for a standard definition of what can be consumed for weight maintenance by all people. However with the availability of the Cambridge Diet the process of trial and error can be a more relaxed procedure, for it is possible to substitute the Cambridge Diet as the sole source of nutrition at any stage in the knowledge that this will lead to a reduction to the target weight with the provision of adequate essential nutrients.

CAMBRIDGE PRODUCTS TO ASSIST AT THE MAINTENANCE STAGE

The aim of the Cambridge plan is to convert people to sensible eating habits, to maintain their target weight and to avoid those factors which are known to be factors in the genesis of disease. However, in the intermediate stage a careful control of the calorie intake is important to determine the level of intake which can be consumed while still retaining the target weight. This would imply calorie calculation and measurement, which is both time-consuming and annoying.

In order to avoid this problem Cambridge Nutrition has produced a set of 'Slim Meals'. These are chicken curry, beef curry, chilli con carne and bolognese sauce with minced beef. While these are meat products, the fat content is very low (between 5.2 g and 9.9 g per meal). When cooked with rice or pasta each meal contains only 400 calories.

RULES FOR SENSIBLE EATING DURING THE MAINTENANCE PROGRAMME STAGE

The patient becomes overweight because more calories are consumed than the body requires. Thus to maintain the newly found set point permanent changes must be made to eating habits. The Cambridge Diet, eaten three time daily during the inititial maintenance period, will provide 330 calories. This allows around 1500 calories for additional food.

Patients can, using a calculator, add up the calorie content of each meal. For most people it is time-consuming and boring, which is why normal food-based weight-reducing diets are not adhered to. On the other hand meals can be planned according to broad principles, avoiding foods of high calorie value. In this respect the following may be helpful advice to patients:

Avoid fat

Fats and oils contain two and a quarter times more calories than protein or carbohydrate, and the quantities of these articles of food should be restricted. This implies that people should avoid fried foods, and grill, poach or boil instead. If it is necessary to fry, a non-stick pan should be used. Lean cuts of meat should be used and any fat removed. Chicken and white fish are advised.

The amount of butter or margarine on bread should be reduced,

and pastry, fatty croissants and cakes, which often contain both fat and sugar, should be avoided.

Butter should not be put on cooked vegetables, nor should mayonnaise or normal salad cream be put on salads. A vinegar or lemon juice dressing with herbs and spices is advised, or a low-calorie salad dressing can be used.

Skimmed milk and low fat yoghurt and cheese should be used in place of full-cream milk and milk products.

Nuts should be eaten only rarely, and then in small quantity.

Restrict sugar

Whenever possible sugar should be substituted by calorie-free sweeteners, such as saccharin or aspartame. The latter is excellent and indistinguishable from sugar with no after-taste. Since most of these sweeteners are unstable to heat they should be added after cooking and not before. Low-calorie drinks should be used in place of cordials, minerals or colas.

Eat more fibre

Fibre can be taken as fruit, vegetables, wholemeal bread or special crispbreads and crisps containing bran. These are healthy, filling, and curb the appetite.

Restrict salt

Excess salt can cause high blood pressure and fluid retention. If it is necessary to add salt, substitute one of the new mineral salts (containing some potassium salts) in place of normal salt.

Restrict alcohol

Alcohol contains one and three quarter times the calories of either carbohydrates or protein, and hence should be drunk only in moderation. Mineral water with ice and a slice of lemon makes an excellent drink for social occasions.

MAINTENANCE OF WEIGHT AND THE SET POINT

A major problem in the maintenance of a reduced weight is the natural tendency for the weight to stabilize upon a 'set point'. Hence as the

weight is reduced by dieting a greater difficulty is experienced. Once dieting is complete and normal eating is resumed the weight rises again to approximately the same level as before.

The set point appears to depend principally on a combination of the metabolic rate and the appetite. As is explained in the chapter on the pathophysiology (Chapter 2), when people go on a slimming diet and reduce their weight, by any form of diet and at any speed of weight reduction, the metabolic rate falls. Unless the calorie intake on the weight-reducing diet is pitched sufficiently below the balance level, the falling metabolic rate will inevitably mean that the energy deficit is reduced as the weight falls, and in consequence the rate of further weight fall is slowed or a plateau occurs.

It is the downward change in the metabolic rate which also explains why dieters put on weight again so easily after resuming their normal food consumption. With the reduced metabolic rate after dieting an increase of calorie intake to even apparently reasonable levels provides more than sufficient energy to supply the needs of the body, and the excess produces a weight increase.

Appetite control also involves various factors, some mechanical (i.e. stomach volume), some chemical (probably at hypothalamic level), some psychological. The extent of the problem of obesity both in human beings and domesticated animals confirms that appetite is not directly related to metabolic needs.

However, the exact control for the metabolic rate/appetite interaction and the set point is very complex, and is currently the subject of a great deal of research. It appears to be related to feedback mechanisms within the endocrine system and its control from the hypothalamic centres. Little can be done currently about weight control by the alteration of the set point. Corrective action can, however, be taken via dietary means. Ultimately the set point appears to be capable of being changed by consistently holding the weight constant at a new level for a long period.

THE LONG-TERM RESULTS WITH THE CAMBRIDGE DIET

In various portions of this book we have referred to the 'yo-yo weight syndrome', and stressed that it is rare for the target weight which is achieved during a dieting process to be maintained over the longer term. For most methods of weight reduction it is accepted that about 95% of those who have lost weight have returned to their old weight within 1 year of the end of the weight-reduction process. The only exception to this has been the long-term weight loss been achieved by surgical bypass procedures, but the widespread use of these is precluded by their high mortality and morbidity (page 30).

It is therefore important to determine the long-term effects that have been achieved by the use of the Cambridge Diet, and particularly by the Cambridge Diet maintenance plan, i.e. the use of the Cambridge Diet for long-term weight maintenance after target weight has been achieved.

In a study in California untertaken by the independent Opinion Research Corporation[61], 661 people who had lost at least 50 lb on the Cambridge Diet were interviewed. The average weight loss in the group was just over 60 lb, and the group represented those with a major weight problem. Even at the stage of the initial approach, the average weight was just over 189 lb with only one in twelve weighing less than 135 lb. Thus at the stage when the survey was started the respondents had lost just over a quarter of their original body weight.

The Cambridge Diet had been used for an average of about 18 weeks at the start of the survey, and less than one in six had been dieting for as much as 6 months prior to the start. Of this period the Diet had been used for about half the time as the sole nutritional source on average. For the rest of the time other foods were also being consumed.

During the first year of the follow-up period the average use of the Diet as the sole source of nutrition was only 5 weeks, though during a further 26 weeks, on average, the Diet was used as part of the total food intake.

At the end of the first year 59% reported that their weight was either the same or lower than that at the start of the survey, and only 41% had shown any rise in their weight. Of those who showed a rise, the average weight gain was 20.3 lb compared to the previous fall in this group of 59.7 lb, i.e. still a net loss of 39.4 lb. A small percentage (7.5%) reported that they were still gaining weight at the end of the year, but even these still had a net loss of 32 lb.

A further follow-up study was undertaken by the Opinion Research corporation 1 year later, i.e. 2 years after the start of the survey[62]. Of the 661 who were interviewed at the end of the first year, 420 could be contacted and interviewed again at the end of the second year. This second year sample is more selected than that at the end of the first year, but it appears to be reasonably representative of the group as a whole.

Of the 420 respondents, 40% had kept their weight either at or below their weight at the end of the first year of the follow-up. Just over a third had not used the Cambridge Diet during the second year. Of those who had used the Diet, the average use was about 20 weeks, of which about 3 weeks was as sole nutritional source.

The average weight gain across the group of 420 respondents was 24 lb; 28 lb for those who had not used the Diet at all during the year

and less than 21 lb for those who had used the Diet. Seventy-four per cent of those who reported a weight gain reported that their weight had now been stabilized. As might be expected, there was a positive relationship between the number of weeks of use of the Cambridge Diet during the past year and the extent to which the weight loss had been maintained. Thus those who had not used the Diet during the past year had retained on average 67% of their original weight loss; those who had used the Diet up to 15 weeks had retained 79%; between 15 and 50 weeks use had retained 93% and those who had used the Diet during over 50 weeks of the second year had retained 98% of the original weight loss (Figure 9).

These results are in stark contrast to the classical previous weight maintenance report[63] in which after 3 years only 5% were successful in retaining their original weight loss.

The study suggests that the Cambridge Nutrition aim of re-educating people to retain their target weight by sensible eating has been only partially successful, but that by the judicious intermittent use of the Diet it is possible for a good weight maintenance programme to be achieved.

EXERCISE FOR MAINTENANCE PURPOSES

In another portion of the book (page 27) we have commented on the possibility of using exercise for the purpose of weight reduction. We have indicated that while the use of exercise, when it is possible within the other medical conditions, is of general benefit, and also increases the weight loss, exercise by itself is not a very efficient means of producing weight reduction.

However, at the maintenance stage, exercise within the competence and potential of the individual is of great merit for the development of a pattern of healthy living.

Exercise can be divided into two types: aerobic and anaerobic. When exercise is aerobic, large quantities of oxygen are being consumed by the muscles, and fat rather than carbohydrate is the main source of the energy. Hence aerobic exercises have the great advantage, even during weight maintenance, of keeping the amount of fat under control.

Aerobic exercises are normally defined as any strenuous continuous use of the muscles for at least 12 minutes – thus for example, swimming, cycling, jogging, walking (for 45 minutes), roller and ice skating are all aerobic providing they they are undertaken on a regular basis.

Obviously in addition to improvement of the muscle tone, such exercise also produces benefit within the cardiovascular and respiratory systems and reduces stress. In order to have any benefit such exercise

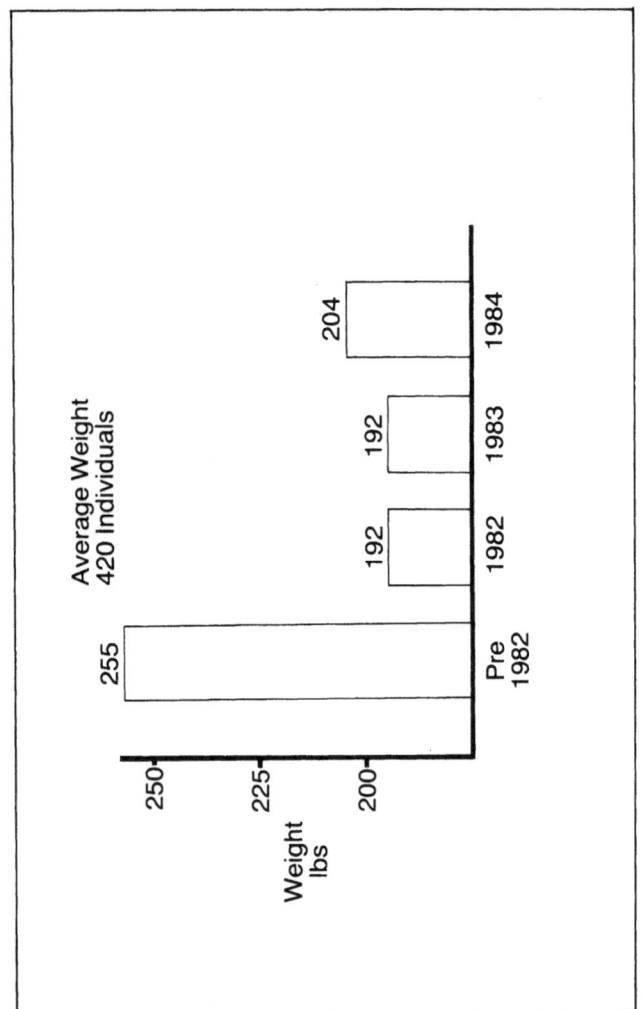

Figure 9 The average weight each year of a group of 420 respondents studied over a 3-year period. (Based on data in the Opinion Research Corporation report[62])

should be undertaken at least three times per week, or preferably every day for a period of at least 12 minutes on each occasion. The exact details of the exercises are clearly outside the remit of this book.

10

The Cambridge Diet as a source of nutrition – current situation and future prospects

The Cambridge Diet was primarily designed for weight reduction. It was based on the concept that the full nutritional needs should be provided within the lowest possible number of calories to avoid the problems inherent in starvation.

The fact that all the nutritional needs – other than calories and fibre (but note that with the use of the nutritional bar the fibre requirement is met) – are contained within a small calorie intake means that the Cambridge Diet can also be used as a good source for these needs other than in the weight reduction situation. When used in this way the Cambridge Diet can be added to the existing food intake, increasing the calorie intake by 330 calories per day (380 calories per day with one meal bar) and ensuring adequate intake of all the essential ingredients. Alternatively, part or all of the normal food intake can be substituted by the Cambridge Diet.

EXISTING NUTRITIONAL USES FOR THE CAMBRIDGE DIET

As supplementary nutrition

Among the groups in which there is anecdotal evidence of nutritional improvement by the use of the Cambridge Diet as a food supplement are the following:

1. in the elderly, particularly those who for various reasons are on an inadequate diet;
2. in pregnancy;
3. during convalescence;

4. in patients who either due to their disease or the treatment they have received find it difficult to swallow solid food.

In some of these people we have advised mixing the Diet with milk rather than water and supplementing the calorie intake with eggs, cream or ice cream for a short period. All these items are high in calories and can be swallowed easily. The high fat content does not in our opinion matter over a short period in these particular circumstances.

Another group who find the Cambridge Diet to be valuable (particularly the meal bars) as a food supplement are those indulging in 'outward-bound' activities. By using the Diet, easily transported rations of high calorie content but poor nutritional value can be used as the main calorie intake, secure in the knowledge that the nutritional needs of the body are being met.

Aid to the prophylaxis of atherosclerosis

It is abundantly clear that atherosclerosis is a multifactorial disorder. Nevertheless it is now widely accepted that animal fats, including dairy products and meat foods, etc., raise the level of cholesterol in the body. Recent epidemiological studies have confirmed the view that has been held for some years, that an increase in the cholesterol level, and particularly in the LDL cholesterol level, is associated with an increased risk of atherosclerosis and coronary heart disease. For this reason there is a widely held view among nutritionists and health experts in the United Kingdom and in other countries, that those people who are potentially susceptible to atherosclerosis should keep the proportion of animal fats to a low level, perhaps replacing them in part by the polyunsaturated fatty acids which appear to have some protective influence in atherosclerosis.

In industrialized countries the consumption of total fat represents about 40% of all the calories that are eaten. Several national public health authorities are now recommending that the proportion should be cut to 30%, and some suggest an even lower level. At the same time they suggest that the ratio between polyunsaturated fat and saturated fat should be increased. This can be achieved by the reduction in animal fats and their partial, but not complete, substitution by vegetable oils.

The Cambridge Diet contains no animal fats. It is very low in total fat (only 3 g/day) of which two-thirds is polyunsaturated. The Cambridge Diet is also cholesterol-free, and although many people believe that dietary cholesterol is not important in the genesis of hypercholesterolaemia, because the body can itself produce choles-

terol, most people would accept that foods rich in cholesterol should be avoided.

Although the dietary fat hypothesis as an important factor in the atherosclerosis risk is widely accepted, others believe that a deficiency of copper or low ratio of copper to zinc in the diet may be an important causative factor. There is much to support this view. For example in a recent survey in the Netherlands it was found that the tissues of people dying with extensive coronary atherosclerosis had a lower copper content than those with minimal disease.

The intake of copper in westernized populations is rather low, and well below the 2 mg per day which is recommended by nutritionists. The Cambridge Diet contains the recommended daily allowance. Should it be true that lack of copper is a contributory fact towards coronary heart disease, the inclusion of the Cambridge Diet in the daily food intake would be an important preventative measure.

Studies with the Cambridge Diet have indicated that, when used as sole source of nutrition, the serum cholesterol level is decreased by an average of about 25% (page 46). All fractions are decreased by about the same proportion. Triglycerides, which may also be a minor risk factor in coronary disease, are also reduced on the Cambridge Diet – by an average of about 40%. Studies are also in progress on the effect of the Cambridge Diet on coagulation factors which may be important in the genesis of arterial occlusion. The preliminary results appear to be encouraging.

Clearly the Cambridge Diet cannot be used as sole source for prolonged periods in those with coronary artery disease whose body weight is normal, although the loss of weight produced by the Cambridge Diet is also a valuable factor in the reduction of coronary disease risk in the obese.

When the patient has a normal weight the Cambridge Diet may be used as a valuable basis for the designation of a good diet for those at risk of coronary artery disease by adding appropriate additional nutrients to raise the calorie value to that which is required from the 330 calories provided by the Cambridge Diet.

Within the additional foodstuffs animal fats should be maintained at as low a level as possible, and vegetable oil should be used when a fat source is needed – for example in cooking. An additional factor which many nutritionists believe is important is dietary fibre, present in wholemeal flour, fruit and vegetables. The Cambridge Diet (except for the meal bar), for reasons of palatability, does not contain very much fibre, and when used as a nutritional base for those at risk of coronary disease it is important that the amount of fibre be increased. This can be done most easily by taking one of the diet meals in the form of the meal bar, and by the consumption of wholewheat products,

including breakfast cereals and breads, together with plenty of fresh fruit and fibrous vegetables.

Hypertension

There is clear evidence that the incidence of hypertension in the population is increasing. Hypertension is a significant factor in the morbidity of the population, particularly as a risk factor in cerebrovascular accidents. It is also a factor in the genesis of coronary disease.

While hypertension can now be kept under control by anti-hypertensive drugs, many of these produce undesirable side-effects, and it is therefore more logical to try to reduce the incidence of hypertension itself. It is now widely agreed that the main nutritional factors contributing to hypertension are:

1. too much salt (sodium) in the diet,
2. not enough potassium,
3. inadequate levels of polyunsaturated fatty acids,
4. excessive alcohol intake.

It has been suggested that the current high incidence of hypertension, predominantly in industrialized countries, could be reduced if the sodium intake were reduced drastically. Health authorities recommend that the current intake of about 12 g per day should be reduced to about 4 g per day. Potassium, on the other hand, which counteracts the sodium effect, could with advantage be increased.

The Cambridge Diet could make an important contribution in this area if it is used as a component of the total food intake in hypertensive people. It contains only 1.5 g of sodium per day, but is higher in potassium (2 g per day) thus providing a moderately low-sodium and high-potassium diet. Although the total quantity of polyunsaturated fats is not high, two-thirds of the fat content of the Cambridge Diet (equal to 2 g per day) is in the polyunsturated form. Hence, three of the four factors are met by the Cambridge Diet.

When the Cambridge Diet is used as sole source the blood pressure falls. Thus Kreitzman et al.[43] found that there was a substantial reduction in the blood pressure of hypertensives. The blood pressure appears to remain lower than its previous value even during the maintenance stage after full weight reduction has been achieved.

For the prevention or relief of hypertension, when the patient is at a normal weight, it is recommended that the Cambridge Diet should be used as a basis for the food intake, and that the additional calories to a weight control level should be provided by foods that are low in sodium, relatively rich in polyunsaturated fats for any fat intake, and high in fibre. Alcohol intake should be kept to a reasonable level.

Adult diabetes

The risk of type II diabetes is very high in overweight people. In fact the majority of people who become diabetic when adults are overweight, and weight reduction is an important part of the treatment of type II diabetes.

Diabetics, like most other overweight people, find it difficult to lose weight and maintain their weight loss on conventional diets. The introduction of the Cambridge Diet has made it much easier for diabetics to lose weight and to maintain their target weight. This weight loss has been accompanied by an improvement in the diabetic state and oral antidiabetic agents which had previously been used may often be discontinued.

Thus the Mancini group in Naples[44] took a number of overweight mild type II diabetics and put them on the Cambridge Diet as their sole form of treatment for 6 weeks. After a few days their blood sugar became normal and the associated high blood fats (both cholesterol and triglycerides) also decreased. These beneficial changes persisted to the end of the study. Earle[64] has confirmed that the Cambridge Diet improves the status of type II diabetics.

Apart from the weight reduction that is possible with the Cambridge Diet, and which leads to the improvement in the diabetic state, the Diet can also be used as part of the nutritional intake for the maintenance of the target weight and hence the reduction in the need for oral anti-diabetic compounds for type II diabetics.

Standard principles for the selection of the additional foods to bring the calorie intake to the required level should be followed. Dietary fibre appears to be valuable as a component in the food intake of such patients.

Athletes

There is now considerable evidence that athletic performance is improved by a good nutritional state, and particularly that poor nutrition is associated with poor performance.

It is abundantly clear that the Cambridge Diet can be used for appropriate weight reduction in athletes who are either overweight for prolonged periods, or those who become overweight as a result of an injury or close-season layoff.

When used for this purpose the Cambridge Diet can be used as the sole food source, although hard physical exercise should not be advised during the first few days on sole source nutrition. There are great merits in using the Cambridge Diet with a 400-calorie meal once a day in weight reduction in athletes, for this enables both a reasonable rate

of weight loss and adequate calorie intake to enable training to be undertaken at the same time.

Once the ideal target weight has been achieved the Cambridge Diet can be used with benefit for maintaining the nutritional state both during the training period and during active performances. For some sports, where weight increase is not a disadvantage, it is convenient to add the Cambridge Diet to the normal food intake. For those sports where weight must be kept low the target weight may be maintained by the substitution of an appropriate quantity of food by the normal intake of Cambridge Diet.

FUTURE PROSPECTS

Multiple sclerosis

During the early use of the Cambridge Diet in the United States, some patients with multiple sclerosis used the Diet for the reduction in the obesity to which their lack of mobility makes such people particularly susceptible. These patients found that coincident with their weight reduction there was an improvement in some of the symptoms associated with their multiple sclerosis.

To date some 30 patients in the United States, and 50 patients in the United Kingdom, suffering from multiple sclerosis are known to be taking the Diet, sometimes initially as sole source for weight reduction, or for those at the target weight, as a nutritional source.

Multiple sclerosis is a disorder in which it is notoriously difficult to determine whether benefit has occurred, for the natural history is that of remissions and relapses at irregular intervals and for variable lengths of time. Certainly the pilot studies to date do not allow any definitive comments to be made about the possible value of the Cambridge Diet in multiple sclerosis other than the fact that weight loss can be achieved with no deleterious effects on the underlying disease.

The pilot studies give no indication that the disease process itself is being influenced. On the other hand there is some initial indication that the relapses are less severe, less regular and less protracted. A feature common to many of those who have used the Cambridge Diet to date is that one of their very troublesome symptoms – fatigue – appears to be substantially reduced. As might be anticipated, the best results appear to have been achieved in those who were initially overweight, and the picture is therefore probably complicated by the improved mobility by weight reduction and by the additional attention that these patients have received from Counsellors or physicians.

Hence while no claims can be made for the Diet at present, further controlled studies are being undertaken with a view to confirming these chance observations.

Food allergies

The extent of food allergy as a factor in the production of various disorders is a matter of current controversy. However, it is now accepted that certain diseases do have a food allergy as one component of the disorder.

The Cambridge Diet is, by its nature, relatively low in allergens. It contains milk protein, to which some people are allergic, and some of the flavouring and colouring agents are also allergens in a minimal proportion of people. These colours and flavours have had to be used to produce a palatable product. Their level has been kept to a minimum and the substances have been selected from the EEC list – wherever possible these being replaced by natural or nature-identical ingredients. Among the present flavours, vanilla and banana are the two which should show the least allergenic risk.

In consequence, while it is clearly not possible to say that the Cambridge Diet can be used successfully in all those suffering from food allergy – this clearly will never be so because of the need for a protein component – anecdotal evidence exists that some patients who previously experienced considerable problems due to food allergies are improved when the Cambridge Diet is used as a the sole source of nutrition.

In those with known or suspected food allergies, where it is clear that milk is not the source of the problem, it is suggested that the vanilla-flavoured Cambridge Diet should be used as the basis for food allergen study. If the patient experiences no problems under these circumstances, additional foods can be added, one or more at a time to determine those foods which may be consumed safely.

Skin disorders

Anecdotal evidence exists that certain chronic skin and nail disorders improved coincident with the use of the Cambridge Diet for weight reduction. How far this is a measure of the removal of a skin allergen, and how far it represents the positive feature of complete nutrition, is currently unknown, but further studies are being undertaken to determine the range of skin disorders in which the Cambridge Diet might have value as a nutritional source.

Rheumatic disorders

The increase in weight which results from their lack of mobility, and the subsequent additional strain which this imposes on weight-bearing joints, is one of the problems of those afflicted with rheumatic disorders.

Weight reduction by the use of the Cambridge Diet has been beneficial in such patients.

There is also some evidence that nutritional factors may be one aspect in the genesis of some cases of rheumatoid arthritis. There is anecdotal evidence that this aspect may also be improved by the use of the Cambridge Diet, and further research into this area is contemplated.

11

The Howard Foundation

The Howard Foundation is a charitable trust which one of us (A.N.H.) established in 1982. The royalties and profits from the sale of the Cambridge Diet are paid to the Foundation, which owns Cambridge Nutrition Limited.

The aims of the Foundation are to provide for biomedical research at the University of Cambridge and other universities in England and Wales; the construction and maintenance of buildings at Downing College Cambridge; the establishment of a cultural centre in the City of Cambridge for the promotion of public education and the appreciation of the arts and crafts amongst the inhabitants; and research into the means of relieving malnutrition especially in underdeveloped countries, particularly using the know-how and technology available to the Trustees and Trust Companies associated with the Foundation. Hence all the funds generated by the commercial exploitation of the Diet are channelled back into worthwhile objectives.

Since its inception, the Trust has supported the following bodies:

The University of Cambridge Department of Medicine;
The University of Wales Institute of Science and Technology,
    Department of Applied Biology;
The Imperial College of Science and Technology, London;
Charing Cross Hospital, London;
Downing College, Cambridge.

Among the current areas in which the Foundation is funding research are:

Obesity
Multiple sclerosis
Diabetes
Longevity
Coronary heart disease and atherosclerosis
The relationship of minerals to metabolic disorders.

The Foundation is also planning to launch a journal on the problems of malnutrition in underdeveloped countries.

APPENDIX 1

Articles in medical and scientific journals on the Cambridge Diet (alphabetical order)

Atkinson, R. L., Berke, L. K., Kaiser, D. L., and Pohl, S. L. (1985) Effects of very low calorie diets in glucose tolerance and diabetic mellitus in obese humans. In Blackburn, G. L. and Bray, G. A. (eds.) *Management of Obesity by Severe Caloric Restriction,* pp. 279–280. (Littlejohn, Mass.: PSG Publ. Co.)

Carman, J. M. (1985) An analysis of user experience with a very low calorie diet. In Blackburn, G. L. and Bray, G. A. (eds) *Management of Obesity by Severe Caloric Restriction.* pp. 33–49. (Littlejohn Mass.: PSG Publ. Co.)

Cook, R. F., Howards, A. N., and Mills, I. H. (1981) Low-dose mianserin as adjuvant therapy in obese patients treated by a very low calorie diet. *Int. J. Obesity,* 5, 267–272.

DiBiase, G., Mattioli, P. L., Contaldo, F., and Mancini, M. (1981) A very low calorie formula diet (Cambridge Diet) for the treatment of diabetic-obese patients. *Int. J. Obesity,* 5, 319–324

Grant, A. M., Edwards, O. M., Howards, A. N., Challand, G., Wraight, E. P., and Mills, I. H. (1978) Thyroidal hormone metabolism in obesity during semi-starvation. *Clin. Endocrinol.,* 9, 227–231

Henry, R. R., Scheaffer, L. and Olefsky, J. M. (1985). Glycemic effects of intensive caloric restriction and isocaloric refeeding in non-insulin-dependent diabetes mellitus. *J. Clin. Endocrinol. Metab.,* 61, 917–925

Henry, R. R., Wallace, P., and Olefsky, J. M. (1986) The effects of weight loss on the mechanisms of hyperglycemia in obese noninsulin-dependent diabetes mellitus (In press)

Henry, R. R., Wiest-Kent, T. A., Scheaffer, L., Kolterman, O. G., and Olefsky, J. M. (1986) Metabolic consequences of very low calorie diet therapy in obese noninsulin-dependent and non-diabetic subjects. (In press)

Hickey, N., Daly, L., Bourke, G., and Mulcahy, R., (1981) Outpatient treatment of obesity with a very low calorie formula diet. *Int. J. Obesity,* 5, 227–230

Howard, A. N. (1975) Dietary treatment of obesity. In Silverstone, T. (ed.) *Obesity: Its Pathogenesis and Management.* pp. 123–154. (Lancaster: MTP Press)

Howard, A. N. (1979) The treatment of obesity by starvation and semi-starvation. In Munro, J. F. (ed.) *The Treatment of Obesity.* pp. 139–164. (Lancaster: MTP Press)

Howard, A. N. (1979) Possible complications of long-term dietary treatment of obesity. In Mancini, M., Lewis, B. and Contaldo, F. (eds.) *Proceedings of Serons Symposium.* pp. 349–363. (London: Academic Press)

Howard, A. N., and McLean Baird, I. (1972) The long term treatment of obesity by low calorie semi-synthetic formula diets, IX International Congress of Nutrition, Mexico

APPENDIX 1

Howard, A. N., and McLean Baird, I. (1973) The treatment of obesity by low calorie diets containing amino acids. *Nutrition and Dietetics IRCS* (73–79) 31–12–4

Howard, A. N., and McLean Baird, I. (1974) The treatment of obesity by low calorie semi-synthetic diets. In Howard, A. N. (ed.) *Advances in Obesity Research*. Vol. 1, pp. 270–273. (London: Newman)

Howard, A. N., and McLean Baird, I. (1977) Very low calorie semi-synthetic diets in the treatment of obesity. An inpatient/outpatient study, *Nutr. Metab.*, 21, 59–61

Howard, A. N., and McLean Baird, I. (1977). A long-term evaluation of very low calorie semi-synthetic diets: an inpatient/outpatient study with egg albumin as the protein source. *Int. J. Obesity*, 1, 63–78

Howard, A. N., Grant, A., Edwards, O., Littlewood, E. R., and McLean Baird, I. (1978) The treatment of obesity with a very low calorie liquid-formula diet: an inpatient/outpatient comparison using skimmed-milk protein as the chief protein source. *Int. J. Obesity*, 2, 321–332

Howard, A. N., Grant, A., Challand, G., Wraight, E. P., and Edwards, O. (1977) Thyroid metabolism in obese subjects after a very low calorie diet. Second International Congress on Obesity

Howard, A. N., Grant, A., Challand, G., Wraight, E. P. and Edwards, O. (1978) Thyroid metabolism in obese subjects after a very low calorie diet. *Int. J. Obesity*, 2, 391

Howard, A. N. (1981) The historical development, efficacy and safety of very low calorie diets. *Int. J. Obesity*, 5, 195–208

Howard, A. N., and McLean Baird, I. Physiopathology of protein metabolism in relation to very low calorie regimens. In *Recent Advances in Obesity Research: III*, (London: Libbey)

Howard, A. N. (1984) The Cambridge Diet: a response to criticism. *J. Obesity Weight Regul.* 3, 65–84

Howard, A. N. (1985) Development of a very low calorie diet – an historical perspective. In Blackburn, G. L., and Bray, G. A. (eds) *Management of Obesity by Severe Caloric Restriction*. pp. 3–20. (Littleton, Mass.: PSG Publ. Co.)

Kreitzman, S. N. (1985) Clinical experience with a very low calorie diet. In Blackburn, G. L., and Bray, G. A. (eds.) *Management of Obesity by Severe Caloric Restriction*. pp. 359–367. (Littlejohn, Mass.: PSG Publ. Co.)

Kreitzman, S. N. Low calorie formulated foods for weight reduction. *Cereal Foods World*, 30, 845–847

Kreitzman, S. N., Pedersen, M., Budell, W., Nichols, D., Krissman, P., and Clements, M. (1984) Safety and effectiveness of weight reduction using a very-low-calorie formulated food. *Arch. Intern. Med.*, 144, 747–750

Krotkiewski, M., Toss, L., Björntorp, P., and Holm, G. (1981) The effect of a very low calorie diet with and without chronic exercise on thyroid and sex hormones, plasma proteins, oxygen uptake, insulin and c peptide concentrations in obese women. *Int. J. Obesity*, 5, 287–293

Lamberts, S. W. J., Visser, T. J., and Wilson, J. H. P. (1979) The influence of caloric restrictions on serum prolactin. *Int. J. Obesity*, 3, 75–81

McLean Baird, I., Parsons, R. L., and Howard, A. N. (1974) Clinical and metabolic studies of chemically defined diets in the management of obesity. *Metabolism*, 23, 645–657

McLean Baird, I., and Howard, A. N. (1977) A double-blind trial of mazindol using a very low calorie formula diet. *Int. J. Obesity*, 1, 271–278

McLean Baird, I., Littlewood, I. R., and Howard, A. N. (1977) Faecal transit time and nitrogen balance in patients receiving a new low calorie formula diet. Second International Congress on Obesity

McLean Baird, I., Littlewood, E. R., and Howard, A. N. (1979) Safety of very low calorie diets. *Int. J. Obesity*, 3, 399

THE CAMBRIDGE DIET

McLean Baird, I (1981) Low calorie formula diets – are they safe? *Int. J. Obesity*, 5, 249–256

McLean Baird, I. (1985) Ambulatory monitoring of obese subjects in normal and very low calorie diets. In Blackburn, G. L. and Bray, G. A. (eds.) *Management of Obesity by Severe Caloric Restriction.* pp. 215–222. (Littlejohn, Mass.: PSG Publ. Co.)

McLean Baird, I. (1986) Holter monitoring studies on patients receiving a very low calorie diet. (In press)

Moore, R., Grant, A. M., Howard, A. N., and Mills, I. H. (1980) Treatment of obesity with triiodothyronine and a very-low-calorie liquid formula diet, *Lancet*, 2 Feb., 223–236

Moore, R., Grant, A. M., Howard, A. N., Mehrishi, J. N., and Mills, I. H. (1981) Changes in thyroid hormone levels, kinetics and cell receptors in obese patients treated with T₃ and a very low calorie formula diet. In *Recent Advances in Clinical Nutrition*, (London: Libbey)

Moore, R., Mehrishi, J. N., Verdoorn, C., and Mills, I. H. (1981) The role of T₃ and its receptor in efficient metabolisers receiving very low calorie diets. *Int. J. Obesity*. 5, 283–296

Scheaffer L., Henry R. R., and Olefsky, J. M. (1985) Glycemic effects of intensive calorie restriction and 180 calorie refeeding in non-insulin dependent diabetics. *J. Clin. Endocrinol. Metab.*, 61, 917–925

Shapiro, H. J. (1978) Report of a comparative study: a new very low calorie formula diet versus a conventional diet in the treatment of obesity. *Int. J. Obesity*, 2, 392

Trott, D. C., and Tyler, F. H. (1981) Evaluation of the Cambridge Diet. A new very-low-calorie liquid-formula diet, *West. J. Med.*, 30(1), 18–20

Visser, F. J., Lamberts, S. W. J., Wilson, J. H. P., Docter, R., and Hennemann, G. (1978) Serum thyroid hormone concentrations during prolonged reduction of dietary intake. *Metabolism*, 27(4), 405–409

Wilson, J. H. P., and Lamberts, S. W. J. (1979) Nitrogen balance in obese patients receiving a very low calorie liquid formula diet. *Am. J. Clin. Nutr.*, 32, 1612–1616

Wilson, J. H. P., and Lamberts, S. W. J. (1981) The effect of triiodothyronine on weight loss and nitrogen balance of obese patients on a very low calorie liquid formula diet. *Int. J. Obesity*, 5, 279–282

Wilson, J. H. P., and Lamberts, S. W. J. (1981) The effect of obesity and drastic caloric restriction on serum prolactin and thyroid stimulating hormone. *Int. J. Obesity*, 5, 275–278

APPENDIX 2

Books on the Cambridge Diet

Birch, R. D. (1982) *The Cambridge Diet: Medically speaking,* Hexi Publishing, P.O. Box 560, Sandy, UT 84091

Blanton, B., Goldstein, J. M., and Silverman, A. (1983) *The Cambridge Diet Psychologically Speaking,* 1983. (Washington: Breakthrough Publishing)

Boe, E. (1983) *The Official Diet Book.* (New York: Bantam)

Howard, A.N. (1985) *The Cambridge Diet.* (London: Jonathan Cape)

Ignasias, Sheila and Dennis (1981) *Recipes for use with the Internationally acclaimed Cambridge Diet Plan,* (Salisbury, Maryland: Slim-lines Publications)

Wilson, F. C. (1983) *The Cambridge Miracle,* Atlantis Publishing of Palm Beach Inc., P.O. Box 2096, Boca Reton, Fl. 33432

APPENDIX 3

Other papers in medical and scientific publications on very low calorie diets

Amatruda, J. M., Middle, T. L., Patton, M. L., and Lockwood, D. H. (1983) Vigorous supplementation of a hypocalorie diet prevents cardiac arrhythmias and mineral depletion. *Am. J. Med.*, 74, 1016–1022

Apfelbaum, M., Bagaigts, F., Glachetti, I., and Serog, P. (1981) Effects of a high protein, low energy diet in ambulatory subjects with special reference to nitrogen balance. *Int. J. Obesity*, 5, 117–130

Apfelbaum, M., Bostsarron, J., Brigant, L., and Supin, H. (1967) La composition dipoids diete hydrique. Effects de la supplementation protidique. *Gastroenterologie*, 108, 121–134

Apfelbaum, M., Boudon, P., Lacatis, D., and Nillus, P. (1917–20) Metabolic effects of dietary protein in 41 obese subjects. *Presse Med.*, 78, 44

Askanazi, J., Weissman, C., Armour Forse, R., and Gil, K. M. (1985) Effects of very low calorie diets on respiratory function. In Blackburn, G. L. and Bray, G. A. (eds) *Management of Obesity by Severe Caloric Restriction.* pp. 251–261. (Littleton, Mass.: PSG Publ. Co.)

Atkinson, R. L., and Kaiser, D. L. (1981) Non-physician supervision of a very low calorie diet. Results in over 200 cases. *Int. J. Obesity*, 5, 237–241

Bistrian, B. R., Blackburn, G. L., Flatt, J. P., Sizer, J., Scimshaw, N. S., and Sherman, M. (1976) Nitrogen metabolism and insulin requirements in obese diabetic adults on a protein sparing modified fast. *Diabetes*, 25, 494–504

Bistrian, B. R., Blackburn, G. L., and Stanbury, J. B. (1977) Metabolic aspects of protein sparing modified fast in the dietary management of Prader-Willi obesity. *N. Engl. J. Med.*, 296, 774–779

Bistrian, B. R., Sherman, M., and Young, V. (1981) The mechanisms of nitrogen sparing in fasting supplemented by protein and carbohydrates. *J. Clin. Endocrinol. Metab.*, 53, 874–878

Bistrian, B. R., Winterer, J., Blackburn, G. L., Young, V., and Sherman, M. (1977) Effect of protein sparing diet and brief fast on nitrogen metabolism in mildly obese subjects. *J. Lab. Clin. Med.*, 89, 1030–1035

Blackburn, G. L., Bistrian, R. W., and Flatt, J. P. (1975) Role of protein sparing fast in a comprehensive weight reduction programme. In Howard, A. N. (ed.) *Recent Advances in Obesity Research*, Vol. 1, pp. 279–281. (London: Newman)

Blackburn, G. L., Flatt, J. P., Cloves, G. H. A., O'Donnell, T. F., and Hensle, T. (1973) Protein sparing therapy during periods of starvation with sepsis of trauma. *Ann. Surg.*, 177, 588–593

Blackburn, G. L., Lynch, M. E., and Wong, S. L. (1985) Use of the very low calorie diets in surgical patients. In Blackburn, G. L., and Bray, G. A. (eds) *Management of Obesity by Severe Caloric Restriction.* pp. 281–293. (Littleton, Mass: PSG Publ. Co.)

Bollinger, R. E., Lukert, B. P., Brown, R. V., Guevara, R. W., and Steinberg, R. (1966)

Metabolic balance of obese subjects during fasting. *Arch. Intern. Med.*, 118, 3–8

Calloway, D. H., and Spector, H. (1954) Nitrogen balance as related to caloric and protein intake in active young men. *Am. J. Clin. Nutr.*, 2, 405–411

Contaldo, F., Di Biase, G., Fischetti, A., and Mancini, M. (1981) Evaluation of the safety of very low calorie diets in the treatment of severely obese patients in a metabolic ward. *Int. J. Obesity*, 5, 221–226

Dehaven, J., Sherwin, R., Hendler, R., and Felig, P. (1980) Nitrogen and sodium balance and sympathetic-nervous system activity in obese subjects treated with a low-calorie protein or mixed diet. *N. Engl. J. Med.*, 302, 478–482

de Silva, R. A. (1985) Ionic, catecholamine and dietary effects on cardiac rhythm. In Blackburn, G. L., and Bray, G. A. (eds). *Management of Obesity by Severe Caloric Restriction*. pp. 183–204. (Littleton, Mass.: PSG Publ. Co.)

Dietz, H. H. and Greenberg, I. (1985) Clinical experience with the use of a mulifaceted program including very low calorie diets. In Blackburn, G. L., and Bray, G. A. (eds). *Management of Obesity by Severe Caloric Restriction*. pp. 335–348.

Ditschuneit, H., Wechsler, J. G., and Ditschuneit, H. H. (1985) Clinical experience with a very low calorie diet. In Blackburn, G. L. and Bray, G. A. (eds). *Management of Obesity by Severe Caloric Restriction*. pp. 319–334. (Littleton, Mass.: PSG Publ. Co.)

Drenick, E. J., and Johnson, D. (1978) Weight reduction by fasting and semi-starvation in morbid obesity. Long-term follow-up. *Int. J. Obesity*, 2; 123–132.

Drenick, E. J., Blumfield, D. E., Fisler, J. S. and Lowy, S. (1985) Cardiac function during very low calorie reducing diets with dietary protein of good and poor nutritional quality. In Blackburn, G. L., and Bray, G. A. (eds). *Management of Obesity by Severe Caloric Restriction*. p. 223–234. (Littleton, Mass.: PSG Publ. Co.)

Evans, F. A. (1938) Treatment of obesity with low-calorie diets: report of 121 additional cases. *Int. Clin.*, 3, 19–23

Evans, F. A., and Strang, J. M. (1929) A departure from the usual methods of treating obesity. *Am. J. Med. Sci.*, 177, 339–348

Fisler, J. S., and Drenick, E. J. (1985) Nitrogen economy during very low calorie reducing diets: a comparison of soya and collagen protein supplements. In Blackburn, G. L. and Bray, G. A. (eds) *Management of Obesity by Severe Caloric Restriction*. pp. 83–98. (Littleton, Mass.: PSG Publ. Co.)

Garlick, P. H., Clugston, G. A., and Waterlow, J. C. (1980) Influence of low-energy diets on whole-body protein turnover in obese subjects. *Am. J. Physiol.*, 238, E235–E244

Genuth, S. M. (1979) Supplemented fasting in the treatment of obesity and diabetes. *Am. J. Clin. Nutr.*, 32, 2579–2586

Genuth, S. M. (1985) Perspective on very low calorie diets in the treatment of obesity. In Blackburn, G. L., and Bray, G. A. (eds) *Management of Obesity by Severe Caloric Restriction*. pp. 21–32. (Littleton, Mass.: PSG Publ. Co.)

Genuth, S. M., and Vertes V. (1974) Weight reduction by supplemented fasting. In Howard, A. N. (ed.) *Recent Advances in Obesity Research*. Vol. 1, pp. 277–278. (London: Libbey)

Genuth, S. M., Castro, J. H., and Vertes, V. (1974) Weight reduction in obesity by outpatients semi-starvation. *J. Am. Med. Assoc.*, 230, 987–991

Genuth, S.M., Vertes, V., and Hazelton, J. (1978) Supplemented fasting in the treatment of obesity. In Bray, G. A. (ed.) Recent Advances in Obesity Research. Vol. 2, pp. 370–378. (London: Newham)

Himms-Hagen, J. (1985) Very low energy diets and changes in metabolic rate: role of brown adipose tissue. In Blackburn, G. L., and Bray, G. A. (eds). *Management of Obesity by Severe Caloric Restriction*. (Littleton, Mass.: PSG Publ. Co.)

Hoffer, L. J., Bistrian, B. R., and Blackburn, G. L. (1985) Composition of weight loss resulting from very low calorie protein only and mixed diets. In Blackburn, G. L., and Bray, G. A. (eds) *Management of Obesity by Severe Caloric Restriction*. pp. 63–72. (Littleton, Mass.: PSG Publ. Co.)

Hoffer, L. J., Bistrian, B. R., Young, J. B., Blackburn, G. L., and Matthews, D. E. (1984) Metabolic effects of very low calorie weight reduction diets. *J. Clin. Invest.,* 73, 750–758

Isaacs, A. J., and Parry, P. S. (1984) A clinical assessment of Modifast in U.K. general practice. *Postgrad. Med. J.,* 60, (Suppl. 3), 74–82

Lewis Landsberg, L. and Young, J. B. (1985): Changes in metabolic rate: role of catecholamines and the autonomic nervous system. In Blackburn, G. L. and Bray, G. A. (eds) *Management of Obesity by Severe Caloric Restriction.* pp. 129–142. (Littleton, Mass.: PSG Publ. Co.)

Lockwood, J. M., Biddle, T. L., and Amatruda, D. H. (1985) Cardiac arhythmias with collagen based diets but not with more complete diets. In Blackburn, G. L. and Bray, G. A. (eds.). *Management of Obesity by Severe Caloric Restriction.* pp. 205–214. (Littleton, Mass. PSG Publ. Co.)

Mancini, M., Contaldo, F., Rivellese, A.,Verde, F., and Di Marmo, L. (1975) A practical and safe programme of calorie restriction for the treatment of massive obesity. In Howard, A. N. (ed.) *Recent Advances in Obesity Research.* Vol. 1, pp. 273–276. (London: Newman)

Mancini, M., Di Biase, G., Contaldo, F. *et al.* (1981) Medical complications of severe obesity: importance of treatment by very low calorie diets intermediate and long term effects. *Int. J. Obesity,* 5, 341–352

Phinney, S. D. (1985) The metabolic interaction between very low calorie diets and exercise. In Blackburn, G. L., and Bray, G. A. (eds) *Management of Obesity by Severe Caloric Restriction.* pp. 99–105. (Littleton, Mass.: PSG Publ. Co.)

Phinney, S. D., Bistrian, B. R., Blackburn, G. L. *et al.* (1983) Normal cardiac rhythm during hypocaloric diets of varying carbohydrate content. *Arch. Intern. Med.,* 43, 2258–2261

Sherwin, R. S. (1985) Influence of diet and carbohydrate on thyoid hormone metabolism. In Blackburn, G. L., and Bray, G. A. (eds) *Management of Obesity by Severe Caloric Restriction* pp. 155–163. (Littleton, Mass.: PSG Publ. Co.)

Simeons, A. T. W. (1954) The action of chorionic gonadotrophin in the obese. *Lancet,* 2, 946–947

Strang, J. M., McClugage, H. B., and Evans, F. A. (1930) Further studies in the dietary correction of obesity. *Am. J. Med. Sci.,* 179, 687–694

Strang, J. M., McClugage, H. B., and Evans, F. A. (1931) The nitrogen balance during dietary correction of obesity. *Am. J. Med. Sci.,* 181, 336–349

Tuck, M. L. (1985) The effect of very low calorie diets on blood pressure control in obese subjects. In Blackburn, G. L., and Bray, G. A. (eds) *Management of Obesity by Severe Caloric Restriction.* (Littleton, Mass.: PSG Publ. Co.)

Van Gaal, L. F., Snyders, D., De Leeuw, I. H., and Bekaert, J. L. (1985) Anthropometric and calorimetric evidence for the protein sparing effects of a new protein supplemented low calorie preparation, *Am. J. Clin. Nutr.,* 41, 540–544

Vertes, V. (1985) Clinical experience with a very low calorie diet. In Blackburn, G. L. and Bray, G. A. (eds) *Management of Obesity by Severe Caloric Restriction.* pp. 349–358. (Littleton, Mass.: PSG Publ. Co.)

Vertes, V., Genuth, S. M., and Haselton, J. M. (1977) Supplemented fasting as a large-scale outpatient program, *J. Am. Med. Assoc.,* 238, 2151–2153

Wadden, T. A., Stunkard, A. J., and Brownell, K. D. (1983) Very low calorie diets: their efficacy, safety and future. *Ann. Intern. Med.,* 99, 675–684

Wadden, T. A., Stunkard, A. J., Brownell, K. D., and Day, S. C. (1985) A comparison of two very low calorie diets: protein-sparing-modified-fast versus protein-formula-liquid diet. *Am. J. Clin. Nutr.,* 41, 533–539

Wechsler, J. G., Swobodnik, W., Wenzel, H., Ditschuneit, H. H., and Ditschuneit, H. (1984) Nitrogen balance studies during modified fasting. *Postgrad. Med. J.,* 60, 66–73

Weinsier, R. L., Bacon, J. A., and Birch, R. (1985) Time-calorie displacement diet for weight control: a prospective evaluation of its adequacy for maintaining normal nutrition status. In Blackburn, G. L., and Bray, G. A. (eds) *Management of Obesity by Severe Caloric Restriction.* (Littleton, Mass.: PSG Publ. Co.)

Winterer, J., Bistrain, B. R., Bilmazes, C. *et al.* (1980) Whole body protein turnover, studied with ^{15}N-glycine, and muscle protein breakdown in mildly obese subjects during a protein-sparing diet and a brief total fast. *Metabolism, 29,* 575–581

Yang, M-U. and Van Itallie, T. B. (1985) Nitrogen balance in obese subjects during caloric restriction: result of short-term and long-term studies. In Blackburn, G. L., and Bray, G. A. (eds) *Management of Obesity by Severe Caloric Restriction.* pp. 73–82. (Littleton, Mass.: PSG Publ. Co.)

References

1. Old proverb quoted by Jelliffe, D. B., and Jelliffe, E. F. (1975) Fat babies; prevalence, perils and prevention. *Environ. Child Health Monogr.,* 41, 124–159
2. Committee on Nutritional Anthropometry, National Research Council (1956) Recommendations concerning body measurements for the characterization of nutritional status. *Hum. Biol.,* 28, 111–123
3. Roche, A. F., Siervogel, R. M., Chumlea, W. C., and Webb, P. (1981) Grading body fatness from limited anthropometric data. *Am. J. Clin. Nutr.,* 34, 2831–2838
4. Macdonald, F. C. (1986) Quetelet index as index of obesity. *Lancet,* 1, 1043
5. Braddon, F. E. M., Rogers, B., Wadsworth, M. E. J., and Davies, J. M. C. (1986) Onset of obesity in a 36 year birth cohort study. *Br. Med. J.,* 293, 299–303
6. Jeffery, R. W., Folson, A. R., Luepker, R. V. *et al.* (1984) Prevalence of overweight and weight loss behaviour in a metropolitan adult population: the Minnesota Heart Survey. *Experience,* 74, 349–352
7. Van Sonsbeck, J. L. A. (1985) The Dutch by height and weight, difference in height and under and overweight among adults. *Maandbericht Gezondheidstatist,* 6, 5–18
8. Editorial (1986) Health implications of obesity. *Lancet,* 1, 538
9. Bray, G. A. (1985) Obesity, definition and disadvantages. *Med. J. Aust.,* 142, S2–8
10. Lew, E. A. (1969) *Proceedings,* 11th Int. Cong. of COINTRA, p. 277
11. Foster, W. R., and Burton, B. T. (eds) (1985) Health implications of obesity: NIH consensus development conference. *Ann. Intern. Med.,* 103, 979–1077
12. Keys, A., Aravanis, C., Blackburn, H. *et al.* (1972) Coronary heart disease; overweight and obesity as risk factors. *Ann. Intern. Med.,* 77, 15–27
13. Kannel, W. B., Le Bauer, E. J., Dawlier, T. R., and McNamara, P. M. (1967) Relation of body weight to development of coronary heart disease; the Framingham study. *Circulation,* 35, 734–744
14. Staessen, J., Fagard, R., and Amery, A. (1985) Blood pressure, calorie intake and obesity. In Bulpilt, C. J. (ed.) *Handbook of Hypertension.* Vol. 6: *Epidemiology of Hypertension.* (Amsterdam: Elsevier), pp. 131–158
15. Horn, G. (1956): Observations on the aetiology of cholelithiasis. *Br. Med. J.,* 2, 732–737
16. Energy and protein requirements: Report of a Joint FAO/WHO/UNV Expert Consultation. Tech. Rep. Ser. No 724 (1985). WHO, Geneva
17. James, W. P. T., Davies, H. I., Bailes, J., and Dauncey, M. J. (1978) Elevated metabolic rates in obesity. *Lancet,* 1, 1122–1125
18. Prentice, A. M., Black, A. E., Coward, W. A., Davies, H. L., Goldberg, G. R., Morgatroyd, D. R., Ashford, J., Sawyer, M., and Whitehead, R. G. (1986) High energy expenditure in obese women. *Br. Med. J.,* 292, 983–987
19. Stuart, R. B. (1967) Behavioural control of overeating. *Behav. Res. Ther.,* 5, 357–363
20. Evans, F. A., and Strong, J. M. (1929) A departure from the usual methods of treating obesity. *Am. J. Med. Sci.,* 177, 339–343

REFERENCES

21. Strong, J. M., McClugage, H. B., and Evans, F. A. (1931) The nitrogen balance during dietary correction of obesity. *Am. J. Med. Sci.,* 179, 687–693
22. Simeon, A. T. W. (1954). The action of chorionic gonadotrophin in the obese. *Lancet,* 2, 946–948
23. Runcie, J., and Hilditch, T. E. (1974) Energy provision, tissue utilisation and weight loss in prolonged starvation. *Br. Med.J.,* 2, 352–354
24. Drenick, E. J. (1967) Weight reduction with low calorie diets. *J. Am. Med. Assoc.,* 202, 118–124
25. Spaulding, S. W., Chopra, I., Sherwin, R. S., and Lyall, S. S. (1976) Effect of caloric restriction and dietary composition on serum T_3 and reverse T_3 in man. *J. Clin. Endocrinol.,* 42, 197–203
26. Bollinger, R. E., Lukert, B. P., Brown, R. V. *et al.* (1966) Metabolic balance of obese subjects during fasting. *Arch. Intern. Med.,* 188, 3–8
27. Blackburn, G. L., Bistriam, R. W., and Flatt, J. P. (1975) Role of protein sparing fast in a comprehensive weight reduction programme. In Howard, A. N. (ed.). *Recent Advances in Obesity Research.* Vol. 1. (London: Newman), pp. 279–281
28. Blackburn, G. L., Flatt, J. P., Cloves, G. H. A., O'Donnell, T. F., and Di Marmo, L. (1975) Protein sparing therapy during periods of starvation with sepsis or trauma. *Ann. Surg.,* 177, 588–593
29. Editorial (1977) Details released at deaths of ten on liquid protein diets. *J. Am. Med. Assoc.,* 238, 2680–2681
30. Scoville, B. A. (1973) Review of amphetamine-like drugs by the Food & Drug Administration. In Paray, G. A. (ed.). *Obesity in Perspective.* DHEW Publ. No NIH 75–707, p. 441
31. Douglas, J. G., Preston, P. G., Gough, J. *et al.* (1983) Long-term efficacy of fenfluramine treatment of obesity. *Lancet,* 1, 384
32. Bjorntarp, P., de Jounge, K., Krotkicwski, M., Sullivan, L., Sjostrom, L., and Stenberg, J. (1973) Physical training in human obesity. III. Effects of long-term physical training on body composition. *Metabolism,* 22, 1468–1481
33. Hollingsworth, D. R., Amatrude, T. T., and Scheig, R. (1970) Quantitative and qualitative effects of L-triiodothyronine in massive obesity. *Metabolism,* 19, 934–945
34. Mason, E. E., Printen, K. J., Hartford, C. E., and Boyd, W. C. (1975) Optimising results of gastric bypass. *Ann. Surg.,* 182, 405–422
35. Bray, G. A., Greenway, F. L., Barry, R. E., Bentfield,J. R., Fiser, R. L., Dahms, W. T., Arkinson, R. L., and Schwartz, A. A. (1977) Surgical treatment of obesity: a review of our experience and an analysis of published reports. *Int. J. Obesity,* 1, 331–347
36. Howard, A. N., and McLean Baird, I. (1973) The treatment of obesity by low-calorie diets containing carbohydrates. *Int. Res. Comm. System,* 73–9
37. Howard, A. N., and McLean Baird, I. (1977) A long-term evaluation of very low calorie semi-synthetic diets: an in-patient/out-patient study with egg albumin as the protein source. *Int. J. Obesity,* 1, 63–78
38. McLean Baird, I., and Howard, A. N. (1977) A double blind trial of mazindol using a very low calorie formula diet. *Int. J. Obesity,* 1, 271–278
39. Howard, A. N., Grant, A., Edwards, O., Littlewood, E. R., and McLean Baird, I. (1978) The treatment of obesity with a very low calorie liquid formula diet: an in-patient/out-patient comparison using skimmed milk protein as the chief protein source. *Int. J. Obesity,* 2, 321–332
40. Kreitzmann, S. (1986). Personal Communication
41. Hickey, N., Daly, L., Bourke, G., and Mulcahy, R. (1981) Out-patient treatment of obesity with a very low calorie formula diet. *Int. J. Obesity,* 5, 227–230
42. Howard, A. N. (1981) The historical development, efficacy and safety of very low calorie diets. *Int. J. Obesity,* 5, 195–208
43. Kreitzman, S. N., Pedersen, M., Budell, W., Nichols, D., Kreitzman, P., and Clem-

ents, M. (1984) Safety and effectiveness of weight reduction using a very-low calorie formulated food. *Arch. Intern. Med.,* **144**, 747–750

44. DiBiase, G., Mattioli, P. L., Contaldo, F., Mancini, M. (1981) A very low calorie formula diet (Cambridge Diet) for the treatment of diabetic-obese patients. *Int. J. Obesity,* **5**, 319–324

45. Moore, R., Grant, A. M., Howard, A. N., and Mills, I. H. (1980) Treatment of obesity with triiodothyronine and a very low calorie liquid formula diet. *Lancet,* **1**, 223–226

46. FDA 'Talk paper' (1982). Cambridge Diet Update: Dept. of Health, Education & Welfare, T82–95

47. White, P. L. (1982) Nutrition and the new health awareness. *J. Am. Med. Assoc.,* **247**, 2914–2916

48. Malone, J. (1983) Is the Cambridge Diet safe? *Richard Times Despatch,* 2 March

49. Editorial (1981) The ultra-low calorie diet revisited. *Obesity/Bariatric Med.,* **1**, 20

50. James, W. P. T. (1982) A Cambridge Doctor on the Cambridge Diet. *Vegetarian Times,* October

51. Thorson, B. (1981) Cambridge Diet. Co-Operative Extension Service. Utah State University, June *Newsletter*

52. Salt Lake City County Health Department (1981) 'This Diet requires Caution'. quoted by *Desert News,* 10–11 June

53. Mullarkey, B. (1982): The Cambridge Diet Caution. *Vegetarian Times,* September

54. Sours, H. E., Frattali, V. P., Brand, C. D., Feldman, R. A., Forbes, A. C., Swanson, R. C., and Paris, A. C. (1981) Sudden death associated with very low calorie weight regimes. *Am. J. Clin. Nutr.,* **34**, 453–461

55. Lantigua, R. A., Amatruda, J. M., Biddle, T. L., Forbes, G. B., and Lockwood, D. H. (1980) Cardiac arrythmias associated with a liquid protein diet for the treatment of obesity. *N. Engl. J. Med.,* **303**, 735–738

56. McLean Baird, I. (1981) Low calorie diets – are they safe? *Int. J. Obesity,* **5**, 249–256

57. Wadden, T. A., Stunkard, A. J., Bromell, K. D., and Van Itallie, T. B. (1983) The Cambridge Diet: More Mayhem? *J. Am. Med. Assoc.,* **250**, 2833–2834

58. Howard, A. (1984) The Cambridge Diet. *J. Am. Med. Assoc.,* **252**, 898

59. Mclean Baird, I., and Howard, A. N. (1986) Personal Communications

60. Calloway, D. H., and Spector, H. (1854) Nitrogen balance as related to caloric and protein intake in active young men. *Am. J. Clin. Nutr.,* **2**, 405–411

61. Opinion Research Corporation (1983) *Current status of people reporting major weight loss as a result of using the Cambridge Plan Diet: one year later.* San Francisco

62. Opinion Research Corporation (1984): *Current status of people reporting major weight loss as a result of using the Cambridge Plan Diet: two years later.* San Francisco

63. Stunkard, A. J., and McLaren-Hume, M. (1959) The result of treatment for obesity. *Arch. Intern. Med.,* **103**, 79–86

64. Earle, K. (1986). Personal communication

Index